Environmental Management Systems

Environmental Management Systems

B. W. Marguglio

Marcel Dekker, Inc.   New York • Basel • Hong Kong

ASQC Quality Press   Milwaukee

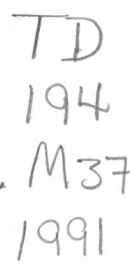

Library of Congress Cataloging-in-Publication Data

Marguglio, B. W.
  Environmental management systems / B. W. Marguglio.
    p.   cm.
  Includes bibliographical references and index.
  ISBN 0-8247-8523-1 (alk. paper)
  1. Industry—Environmental aspects. 2. Industrial management.
I. Title.
TD194.M37   1991
658.4'08—dc20   91-7755
                                                         CIP

This book is printed on acid-free paper.

Copyright © 1991 by ASQC Quality Press
Neither this book nor any part may be reproduced or transmitted in any form or by any means, electronic or mechanical, including photocopying, microfilming, and recording, or by any information storage and retrieval system, without permission in writing from the publisher.

ASQC Quality Press
310 West Wisconsin Avenue, Milwaukee, Wisconsin 53203

Marcel Dekker, Inc.
270 Madison Avenue, New York, New York 10016

Current printing (last digit):
10 9 8 7 6 5 4 3 2 1

PRINTED IN THE UNITED STATES OF AMERICA

Preface

This book fills a void in the environmental literature. There are books on environmental management from a global or regional perspective. There are books on environmental sciences and engineering. There are a few books that touch only lightly on environmental management issues at the plant level. This is not a criticism of these books; it's just that they have orientations different from those of management systems. This book provides operations managers; environmental managers, scientists, and engineers; and environmental regulators (and those who have a public interest) with the specific features of environmental policies and procedures that should exist at the plant level. These features are universally applicable from plant to plant regardless of the nature of the industry or the details of technical requirements. Utility, petrochemical, chemical, pharmaceutical, and other manufacturing industries can benefit technically and economically by implementing the features of the management systems described in this book.

Let me relate a little about how this book came to be written. I was responsible for environmental affairs, along with other functions, at Consumers Power Company, a large electric and gas utility serving most of Michigan. When the company assigned the director of environmental affairs and his staff to report to my office, I had no background in the environmental sciences. At that time, my significant experiences were in project management and quality management. Undoubtedly, the environmental scientists were concerned about my lack of background in their technical arena. So was I.

I am indebted to the environmental scientists who taught me the fundamentals of a variety of environmental sciences. I am also indebted to these scientists for their willingness, in turn, to maintain an open mind toward my management systems and to have thoroughly learned these systems—systems that are addressed in detail in this book. Most important, I am indebted to these scientists for having successfully implemented these systems—success that serves as testimonial to the practicality of this book.

The management systems described in this book were also the subject of environmental management seminars that I led and that were attended by representatives of electric and gas utilities and petrochemical and pharmaceutical manufacturing companies. The value of these management systems was acclaimed in the feedback I received from the attendees.

These management systems are not especially new to industry. Some of the systems have interesting parallels in project management and quality management. However, based on my industrial experiences, my feedback from the seminars and, most significantly, my understanding of the state of environmental management in industry and in regulating agencies, I was compelled to write this book. I hope that it will contribute to a better understanding of the roles required of environmental managers, scientists, regulators, and, by all means, the corporate line and function managers, who play the biggest part in attaining environmental control for safety and for the economic advantage of their companies, their communities, and society as a whole.

I challenge company professionals to compare the management systems that exist in their companies with the breadth and depth of the management systems described in this book. Don't stop there. Make the changes in your companies in order to be compatible with these management systems.

Essentially, the same challenge holds for regulators. Compare the

PREFACE

management systems in this book with the systems in the companies that you regulate and use your special powers to cause these companies to make the management systems improvements described here.

Finally, I urge members of public interest groups to use the information in this book as a baseline with which to assess the environmental control management systems used by your neighboring companies. Progressive companies will cooperate, at least by making their management systems documents available through public libraries.

I appreciate the efforts of many people at Marcel Dekker, Inc., and ASQC Quality Press, copublishers of this book. In particular at Marcel Dekker, Inc., my thanks go to Maria Allegra, Associate Acquisitions Editor; Lila Harris, Supervisor, Book Editorial; Eridania Perez, Executive Director, Promotions and Marketing; Marilyn Ludzki, Copy Supervisor; and Billie Spaight, Senior Copywriter. At ASQC Quality Press, my thanks go to Jeanine Lau, Acquisitions Editor; Tammy Griffin, Production Editor; and Susan Westergard, Marketing Administrator.

B. W. Marguglio

Contents

Preface   *iii*

Chapter 1   Environmental Management Tenets   1

Chapter 2   Environmental Policies   17

Chapter 3   Environmental Requirements Documents   31

Chapter 4   Environmental Considerations in Facility Site Selection   41

Chapter 5   Environmental Considerations During the Design Engineering Process   53

Chapter 6   Permit, License, and Approval Applications   61

Chapter 7   Property Tax Exemptions for Environmental Equipment   83

Chapter 8	Project Management of Environmental Studies	87
Chapter 9	Environmental Agency Inspections	95
Chapter 10	Reporting to Environmental Regulatory Agencies	103
Chapter 11	Environmental Audit	109
Chapter 12	Environmental Noncompliance Reporting and Corrective Action	121
Chapter 13	Contracting for Waste Management Services	135
Chapter 14	Maintenance of Environmental Equipment	157
Chapter 15	Environmental Education and Training	173
Chapter 16	Environmental Awareness and Emergency Response	179

Index 187

Environmental Management Systems

1

Environmental Management Tenets

THE TENETS

There are two tenets underlying the management policies, requirements, and procedures described in this book.

Company managers and environmental regulators have a responsibility to acquire a demonstration of the company's ability to comply with environmental law and commitments.

On a facility-by-facility basis, the company's demonstration must be based on the existence of specific policies, requirements, and procedures for compliance and on the existence of specific assignments of responsibilities for the implementation of these policies, requirements, and procedures.

This demonstration is necessary because the adverse effects of noncompliance may be irreversible and significant relative to the health and safety of the public or the employees, the financial health of the company, or both. Even if the effects are reversible, the cost to the company, its employees, or the "community" is often unacceptably high. There is no need to cite examples of these conditions. Today, the operations of many companies are so massive, the potential adverse effects of these operations so great, and the speed with which these effects may be manifested so rapid, that we have no choice other than to require a before-the-fact demonstration of the company's ability to protect its community, employees, and stockholders. Yet, except for isolated elements of the law, there are no requirements on a universal basis in federal, state, or local laws for such a demonstration.

Before a company is granted a permit to construct a nuclear power plant or granted a license to operate that plant, the law requires such a demonstration. Before a manufacturer is licensed to market a given pharmaceutical product, the law requires such a demonstration. But there is no such preoperational demonstration requirement for petrochemical facilities, fossil-fired power plants, and other similar facilities.

An Environmental Impact Statement (EIS) describes the impact of a facility given that it is designed, constructed, and operated in accordance with its intent. EISs and permits/licenses describe controls that are required during operation. But there must be policies, requirements-type documents, and procedures by which to carry out the design, construction, and operational intent and controls.

Compliance with environmental law is complex because the law stems from a variety of sources and because attainment of compliance with almost any environmental law involves a variety of management and technical functions. How can a company be expected to comply with environmental law without first having established its own policies, requirements, and procedures which are responsive to the law? How can company employees be aware of their responsibilities to the law unless these requirements are interpreted for them and made to be specifically relative to the operations and functions for which they are responsible? How can employees achieve compliance with these requirements and do so economically and consistently unless the methods for achievement are defined, including the interfaces that must exist among the employees in the implementation of these methods? In most companies there exist policies, detailed requirements, and detailed procedures relating to other practices, such as accounting, to mention only one. How then can a company expect to engage in environmental practices or in other practices which have environmental impact, without similar policies, requirements, and procedures?

In the past, companies sought refuge in the so-called safe-harbor defense. The idea behind this defense is that the company's level of culpability is reduced by the absence of any noncompliance with its own policies, requirements, and procedures and, therefore, it is best not to document such things. Not only is this defense on the wane, but its use could result in even stiffer legal penalties for the offending company. Should not a company be more culpable if its failure to comply with the law is caused, in large part, by the absence of any systematic approach to compliance as is provided by documented policies, requirements, and procedures? The absence of a systematic approach is far more offensive because it fosters consistent or repetitive noncompliance. Administrative law judges and the courts now recognize this.

THE LAW

It must be understood that the law is a lot more than legislation alone. A large body of the law is given in the rules and regulations established by federal, state, and local bureaucratic agencies, such as the federal Environmental Protection Agency or Michigan's Department of Natural Resources. These agencies are empowered by the legislation to create the rules and regulations. Often the legislation is on a rather general level. The rules and regulations become the real meat of the law. Although some rules and regulations may be established by the agency with input from industry, commerce, and other private citizenry, sometimes there is no agency obligation to acquire such input. Even when hearings are required through which to obtain outside input, often there is no obligation for the agency to achieve consensus as to the contents of the rules and regulations. In addition, there is often no legislative oversight of the agency's rules and regulations.

Furthermore, some environmental law is established by administrative and judicial rulings. The less specific the environmental legislation and the environmental rules and regulations, the more the need for administrative and judicial rulings. It should also be noted that some of these rulings are made by administrative law judges who are employed by the agency. Their rulings can be overturned only by resorting to the civil courts.

Finally, when a company makes a commitment to an environmental regulator, regardless of whether or not the commitment is made in direct response to environmental law, the commitment may take on the force of law, especially if the commitment is part of a permit or license.

The wide variety of sources of environmental law—namely, environmental legislation, rules and regulations, administrative and judicial rulings and commitments—makes it necessary for the company's environmental experts to convert the law to policies, requirements, and procedures that are well understood by the company's employees.

In the attainment of a single requirement, company personnel from many different organizations may be involved. For example, in order to attain the requirements relative to the amount of sulfur dioxide (SO_2) emission from a coal-fixed power plant a half-dozen or so departments may be involved—namely, the departments responsible for specifying the coal to be procured, for procuring the coal, for inspecting the coal, for operating the plant, for maintaining plant equipment and, of course, the environmental department. The absence of policies, requirements, and procedures in any of these departments can cause the failure to limit emission of SO_2 to the required level. In order to achieve compliance with this regulatory requirement, the company must establish its compliance policy, establish a series of subtier requirements that are related to the SO_2 requirement, assign each of these subtier requirements to a specific organization at the working level, and define the method by which each of these subtier requirements is to be achieved. How can it be otherwise?

TOTAL ENVIRONMENTAL CONTROL

As shown in Figure 1, in order to be reasonably certain of its economic compliance with environmental law and commitments, the company

ENVIRONMENTAL MANAGEMENT·TENETS

Stage of determination	Program elements					
	Adequate design of the management system	Adequate implementation of the management system design	Adequate design of the facility/ equipment system	Adequate implementation of the facility/ equipment system design	Adequate design of the operations & maintenance system	Adequate implementation of the operations & maintenance system design
Prevention of inadequacy	✓	✓	✓	✓	✓	✓
Timely detection and correction of inadequacy	✓	✓	✓	✓	✓	✓

Figure 1 Total environmental control program.

must have designed adequate environmental policies, requirements, and administrative procedures, collectively referred to as the *management system,* and must have consistent implementation of the management system as designed. Of course, an adequately designed and consistently implemented management system alone cannot assure compliance. Its absence, however, will assure noncompliance. The total environmental control program also depends on the adequacy of the design of the facility and its hardware and on the accurate conversion of that design in the construction and installation phase. Once in operation, the total environmental control program depends largely on the adequacy of the design of the operations and maintenance procedures and on the consistency of their implementation. Any inadequacy in the design or inconsistency in the implementation of the management system, facility/equipment, or operating/maintenance procedures will lead to noncompliance.

To begin with, the management system must stress the elements necessary to prevent noncompliance. Recognizing that prevention is not always possible or wholly effective, the management system must also contain elements by which to detect and correct noncompliance.

Relative to any law, the effectiveness of the company's environmental control program is directly related to the "adequacy" of the documented program and the degree of compliance with the program. How can the company economically achieve compliance with the law if its policies, requirements, and procedures are not effective and efficient? What good is compliance with policies, requirements, and procedures if they are inadequate to begin with? In this sense "adequacy" refers to the state of containing nothing more and nothing less than is necessary to get the job done. How can compliance be attained if the management system does not stress prevention to begin with? What good is a system that allows the company to falter even though the condition is subsequently detected—especially recognizing the possible irreversibility of the condition or excessive cost of its correction?

It is important to recognize that the total environmental control program must address itself both to the external environment to be experienced by the community and the internal environment to be experienced by the employees.

ENVIRONMENTAL "MATURITY PROFILE"

A technique that might be used in assessing a company's status relative to the development and implementation of its environmental compliance program is the "maturity profile." This technique was first described by Philip Crosby in his book, *Quality Is Free* (New American Library, 1980), in which he assessed the status of company quality programs. I have adapted his concept for use in assessing environmental programs.

To assess the status of a company's progress relative to environmental compliance, one might consider six major characteristics of the program, as follows:

- Environmental policies
- Environmental requirements
- Environmental procedures
- Environmental compliance measurements and assessments
- Corrective action of environmental program design and implementation inconsistencies
- Organization for environmental compliance.

In addition, one might establish four levels of the company's development, namely:

- Insensitivity to the need for environmental compliance
- Awareness of the need for environmental compliance
- Enlightenment as to the need for environmental compliance
- Certainty that environmental compliance is being attained

Using these six characteristics and four stages of development, the profile shown in Fig. 2 was established.

When a company is at the *insensitivity* stage, there is no established environmental policy. Environmental requirements are communicated only in their original form. Communications are limited. There is no established system by which to assure awareness of changing requirements. There are no measurements or assessments of environmental compliance other than those required by law. Legally required measurements are not used, other than to submit them to a regulator as may be required by law. There are no procedures defining methods for

| | Stage | | | |
Characteristic	Insensitivity	Awareness	Enlightenment	Certainty
Environmental policy				✓
Environmental requirements				✓
Environmental compliance measurement/ assessment				✓
Environmental procedures				✓
Environmental program corrective action				✓
Environmental organization				✓

Figure 2 Environmental compliance "maturity profile."

preventing noncompliance or for detecting and correcting noncompliance. There are no procedures defining other administrative functions supporting the attainment of environmental compliance. There are corrective actions only upon the threat of regulatory action or penalty. The actions correct the immediate noncompliances but not the root causes of the noncompliances. There is no environmental organization. Responsibilities for environmental compliance are not well defined.

At the *awareness* stage, there are established environmental policies, but they fall short of demanding absolute compliance with en-

ENVIRONMENTAL MANAGEMENT TENETS

vironmental laws and other requirements. Environmental laws, in their original forms, are communicated more extensively throughout the organization. The legally required measurements are used to obtain corrective action after the fact of noncompliance. There are some procedures for finding and correcting noncompliance. Generally, there is timely correction of noncompliance. The environmental organization is understaffed and has a staff role only. A staff role, in this sense, means that the organization is authorized only to provide advisory guidance, as contrasted to an organization that has a functional role or a line role.

The effectiveness of an organization that performs a staff role can be increased if the system requires that the organization take part in the review cycle and that the organization's review comments be officially responded to. Of course, the more technically and economically appropriate the organization's comments, the more these comments are sought and the more effective the staff role becomes.

When an organization has a functional role, it has the responsibility and authority to establish criteria by which a certain function will be accomplished. For example, the company's accounting department usually has the responsibility and authority for establishing the criteria for completing and processing individual expense reports. Or, for example, the environmental department can have the responsibility and authority for establishing the points in the process at which environmental measures will be taken, the frequency of those measures, the methods for those measures, and the acceptance criteria.

When an organization has a line role, it has the responsibility and authority for performing one or more steps in the process and for determining the acceptability of the process at any given step, or the appropriateness of moving the process forward beyond any given step. For example, the accounting department reviews completed expense reports and decides whether or not to issue expense reimbursement checks. Likewise, the environmental department may have the responsibility to measure the content of a fluid proposed to be discharged from the plant, decide whether or not the fluid is suitable for discharge, and authorize or prevent its discharge, as appropriate.

When a company has reached the stage of *enlightment,* its environmental policies demand absolute compliance with legal requirements.

There is a system by which to keep abreast of and communicate changing legal requirements—although the requirements are still communicated in their original form. Thus, it is difficult for the responsible organizations to interpret the requirements and for the requirements to be interpreted uniformly by the variety of organizations which may have a role in attaining compliance with any one requirement. At this stage, in addition to the use of legally required measurements to identify the need for and obtain corrective action after the fact, an independent compliance audit system is used to obtain corrective action, also after the fact. There are adequate procedures for detecting and correcting noncompliance. There is formal training in these procedures. There is timely correction of noncompliances. Sometimes, there is even correction of the root causes of the noncompliances. The environmental department is adequately staffed, but still has a staff role only.

At the stage of *certainty,* in addition to absolute compliance with environmental law, environmental policies are established to address issues beyond those covered by the law. The policies also demand the application of risk management to environmental issues that go beyond the law. At this stage, environmental law is converted to specific requirements applicable to specific conditions for each facility. Operating line personnel are specifically trained in these requirements. Operating line personnel accept these requirements. There is a "contract." In addition to legally required measurements and an independent compliance audit system used to obtain corrective action after the fact, at this stage, for each plant or process-specific requirement, there is either an in-process measure or an in-process assessment technique enabling line operating personnel to determine their state of compliance in real or in reasonable time. There is also a compliance self-assessment system that is intended to identify and correct potential problems before the fact. In addition to procedures for detecting and correcting noncompliance, there are procedures for preventing noncompliance. Preventive techniques cover aspects of operations, preventive maintenance, periodic inspection, and administration. Administrative procedures are fully developed. All procedures are specific as to the requirements, responsibilities, and methods. There is consistent correction of noncompliances and of the root causes of the noncompliances—both on a timely basis. Finally, at this stage, the environmental organization has both functional and line roles, as described earlier.

Figure 3 depicts the progress from insensitivity to certainty, as it relates to the frequency of noncompliance. For any given process, at the stage of insensitivity, there are frequent noncompliances and the frequency range is quite wide. Insofar as environmental requirements are concerned, the process is out of control—from the viewpoint that both the absolute frequency of the noncompliances is too high and the variation of the frequency over time is too great.

The most significant reduction in the absolute frequency of environmental noncompliances is achieved in the awareness stage. Still, however, processes that have the potential for adverse environmental impact are out of control.

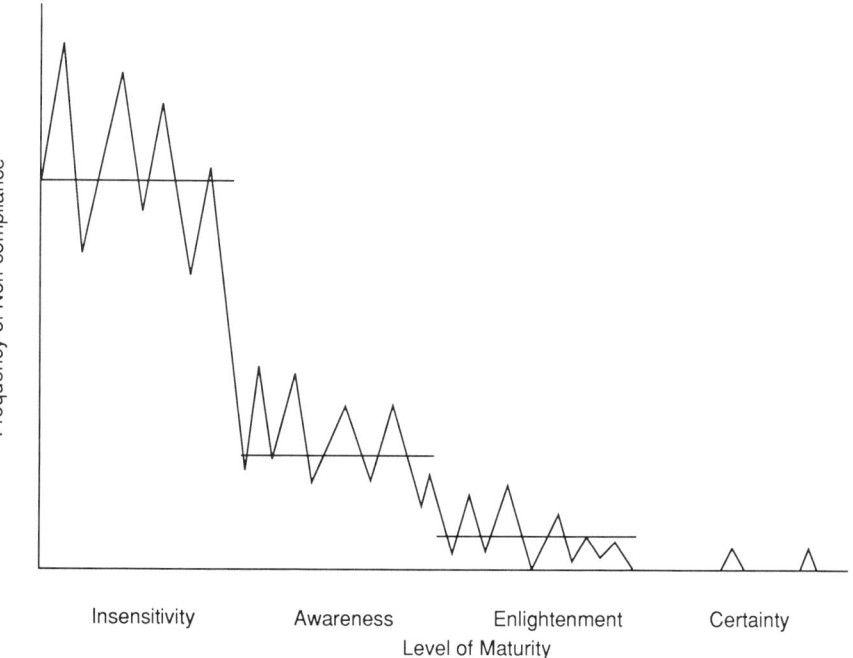

Figure 3 Progress from *insensitivity* to *certainty*.

Enlightenment brings a frequency of noncompliances that enables the company to get by with the regulator and the community. But for the environmental and managerial professional, it is still far from an acceptable level of performance.

Only certainty is acceptable. Certainty is attained through a total environmental program with the major characteristics of that program existing in an advanced state. With certainty comes near absolute compliance and this, in the long term, is the most efficient.

THE MANAGEMENT SYSTEM HIERARCHY

Figure 4 describes the way in which the environmental management system may be communicated in writing. Starting at the top, environmental policies are intended to communicate principles that are universally applicable throughout all activities in which the company is engaged. Examples of environmental policies, along with their rationale, are given in Chapter 2. A policy should be a concrete statement as to the company's desire with regard to a specific aspect of environmental control. The fact that the policy has universal application within the company does not preclude it from having real directional value. A policy should be worded such that anyone who

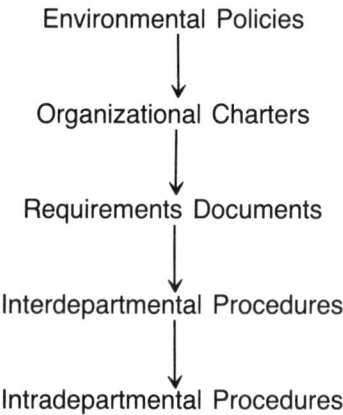

Figure 4 Management system hierarchy.

knows the policy and who exercises reasonable judgment can perform in accordance with the company's intent, even in the absence of specific direction on the subject as may be provided by a procedure. In other words, in following the policy, there should be little chance for an error of significant adverse environmental impact, even if there is a failure to exactly follow the detailed procedure.

In executing environmental policy, each organization has some responsibility and authority as usually described in its organizational charter. The authority assigned to an organization enables it to fulfill its responsibility. An organization should not accept responsibility without corresponding authority. Thus, to the extent that organizational charters describe responsibilities and authorities regarding environmental control, the charters constitute a part of the overall written description of the environmental management system. Chapter 2 also provides examples of environmental department charter responsibilities and authorities.

There may be two levels of procedural documentation that provide the specific design of the environmental management system. Interdepartmental procedures are at the higher level because they provide the requirements, responsibilities, and methods which, for a given system, are applicable to two or more organizations. Interdepartmental procedures particularly emphasize the points of interface among various organizations. The output of a process performed by one organization may serve as the input to a process performed by another organization and the output and input must be compatible. An interdepartmental procedure should be approved by each organization whose activity is affected by the procedure.

Intradepartmental procedures are at the lowest level in the management system hierarchy. They provide the specific requirements, responsibilities, and methods for accomplishing activities which are performed entirely within the organization that issues the procedure.

Starting with Chapter 3, the pages of this book are filled with suggestions relating to the content of interdepartmental and intradepartmental procedures on a variety of environmental control susbjects. Regardless of the level of the procedure (i.e., inter- or intraorganizational) it should be prepared in accordance with the systems approach and good practices.

A question arises regarding the degree of specificity that is appropriate in these environmental management system procedures. There are those who believe that the procedures should be less, rather than more specific to facilitate both flexibility and management understanding. Let us address the flexibility issue first. The opposite of *flexibility* is *inflexibility*. The opposite of *specificity* is *generality*. The degree of specificity or generality of a procedure has nothing to do with its degree of flexibility or inflexibility. Specificity and flexibility are two different concepts. To obtain flexibility, options must be provided. In the absence of options there is no flexibility regardless of whether the procedure is stated in specific or general terms.

The specificity for which there may be concern usually relates to the sequence with which to do the job and how it is to be done. When the "sequence" and "how to" specificity is necessary either for technical adequacy, economic optimization, or consistent interface, the specificity should be part of the procedure regardless of the procedural level in the hierarchy. When the specificity is not necessary to achieve either compliance or efficient operation, the specificity should be omitted from the document so as to allow the individual preference to prevail. Stated differently, if there is a sequence or method by which to better assure that the requirements are being met economically, this sequence or method certainly is appropriate for incorporation into the procedure. If there are various options, the conditions which are prerequisite to the application of each option should be specified as well as the sequence and method for each option when it is technically and economically significant.

The fact that inter- and intraorganizational environmental management system procedures provide sequence and method specificity does not mean that the thinking initiatives of the employees are or have been abrogated. In an atmosphere of participative management, these initiatives must be sought at the time that the procedure is being developed but once developed and implemented the procedure, as issued, must be followed. To do otherwise will lead to disrespect for requirements as a whole. When the sequence and methods affecting compliance and economy are specified, the company receives greater assurance that at least minimum standards will prevail. Subsequently, if conditions should change to warrant a corresponding change in the procedure,

ENVIRONMENTAL MANAGEMENT TENETS 15

the employee initiatives should prevail to identify the need for such change and to assure that the change is made on a timely basis. Of course, management systems are rarely designed to their ultimate perfection; therefore, the initiative to identify potential improvements must always be given to the employee.

Now to the argument that generality is preferable for ease of understanding. This cannot be so by definition. How can one's general knowledge of an activity compare with another's specific knowledge of that same activity. It cannot. Furthermore, it must be understood that procedures are largely for the purpose of educating and training those who do the job on a day-to-day basis. They need these specifics to make their jobs easier and those who manage need these specifics to adequately control.

Finally, when in actuality a process has been made as simple as possible, it cannot be made still simpler merely by describing it in generalities. Oversimplification of the procedure can only lead to performance failure.

2

Environmental Policies

INTRODUCTION

In this chapter a number of typical company environmental policies are presented and discussed. The intent here is to give the reader an insight into a broad range of policies which are of most significance, not to address every conceivable policy.

POLICIES

One of the top level policies of the company should be to

Protect and enhance the environment through planning and prudent use of resources and technology, providing the regulatory agencies and the public with complete and accurate information relative to environmental decisions, including cost information.

The major salient point of this policy is that the environmental protection and enhancement is not necessarily predicated upon the existence of environmental law. In this case, the company recognizes its responsibility to protect and enhance the environment in view of the fact that the company is the technical expert for the process in question. The company need not depend upon the regulator for direction as to what is appropriate for environmental protection and enhancement.

Another salient point is the commitment to environmental enhancement as well as protection. This would be an overcommitment were it not coupled with the commitment to also provide the short and long term costs of enhancement. In other words, the commitment to environmental enhancement is made with the understanding that such enhancement would be limited to those cases for which the cost of the benefits of enhancement exceed the cost of the risks of nonenhancement. The cost estimating techniques for assuring a reasonable trade off can be the subject of a separate discussion.

Of course, having dollar benefits which exceed dollar risks applies to environmental protection as well as to environmental enhancement. When dealing with an issue of enhancement, the company has far more latitude than when dealing with an issue of protection. In the case of protection, the benefit-risk analysis techniques are more subject to overview along with the value systems applied to these techniques and value systems may differ among the company, the regulator, and the public.

The commitment to planning made in this policy is a very important attribute of the policy. How can planning be accomplished in the absence of other policies, requirements, and procedures? Obviously, for complex projects and activities, it cannot.

Finally, this policy commits the company to enable the regulator and the public to deal with the company on a common and equal footing by providing them with complete and accurate information. There should be no real or perceived differences in the data bases with which the company, the regulator, and the public are working. If the differences in data bases are not ferreted out, how can there be an understanding of the differences in value systems among the three parties? Undoubtedly, a lot of problems have gone unresolved or have been concluded unsatisfactorily because of the failure to recognize differences in the factual data bases as contrasted to the differences in the value systems applied to those data bases.

ENVIRONMENTAL POLICIES 19

Another potential environmental policy is that the company shall

Comply with environmental law and other requirements as defined by the environmental department with input from the legal department when a requirement is questionable. Environmental requirements shall address legislation, rules and regulations, administrative and judicial interpretations, company commitments accepted in permits and licenses, and other company commitments.

The most salient point of this policy is that the environmental department is established as the authority for defining environmental requirements for the company, with the understanding that when the legality of the law is questionable or when its interpretation is questionable, the environmental department shall seek input from the legal department, which should be the final authority in interpreting the law.

This policy also implies that the environmental department shall not merely perform a pass-through function—that is, requirements of the law shall be conveyed by the environmental department in terms that relate to the company's business and that are understandable to the company's employees. This is based on the environmental department's expertise in interpreting the technical aspects of the law for the operating organizations, much like a manufacturing engineer interprets the design engineer's drawings and specifications, putting them in terms that can be more easily understood by the fabrication shop personnel.

Another salient point of this policy is that environmental law, as discussed in Chapter 1, stems not merely from legislation, but also from rules and regulations, and administrative and judicial interpretations (which are often overlooked as being a part of the law) and company commitments, which are made as a condition of the granting of a permit or license as well as in the interest of good neighborliness. Permit and license commitments have the force of law and other commitments also often have the force of law. The mention of company commitments that do not stem from a regulatory basis, in and of itself, is significant.

Another policy is that

Operating line organizations shall be responsible primarily for attaining compliance with environmental requirements.

The implications of this policy are very broad. Assuming that the requirements are properly communicated to begin with, this policy means that those who provide the design for the process or who operate the process are held accountable for any failure to meet environmental requirements because of poor design or poor operating practices, respectively. This policy makes it incumbent upon the line organizations to take the actions that are necessary to prevent noncompliance. The line organization remains responsible for compliance attainment even if the environmental department is responsible for measuring or assessing the environmental impact of the process and for feeding back that information to the line organization—in other words, even if the line organization is not making its own measurements or assessments. There is a clear line of demarcation in this policy between the responsibility for attaining compliance and the responsibility for measuring or assessing to determine whether or not compliance has been attained.

Another important policy is that

For each environmental requirement, there shall be a quantitative measurement or a qualitative assessment by which to ascertain the degree of compliance with the requirement, either in real or in reasonable time.

This policy implies that each requirement, to begin with, shall be given quantitatively if possible, and that there shall be a corresponding quantitative measurement. The policy also implies that each qualitative requirement shall be clear and uniformly interpretable in only one way as was intended and that, by virtue of this clarity, the degree of compliance with such a requirement shall be assessable. Furthermore, by this policy, the feedback of the measurement or assessment must be either in real or in reasonable time, which is intended to enable the line organization to use the measure or assessment to prevent noncompliance. If prevention is neither possible nor practical, at least the feedback shall be in time to prevent a significant noncompliance. It should be stressed that in the absence of a measurement or assessment technique, there is no requirement in reality.

ENVIRONMENTAL POLICIES

The following policy is somewhat of a corollary to the preceding policy. This policy is that

For each operating facility or process there shall be a facility- or process-specific document that provides the environmental requirements applicable to the facility, the corresponding measurements and assessments, and the implementing procedures by which to attain compliance with the requirements and by which to measure or assess the degree to which compliance has been attained. The facility-specific document shall be oriented toward preventing noncompliance. The document shall be subject to a change control system.

One new salient point introduced in this policy is that the requirements, measurements and assessments, and procedures shall be specific to the facility or process to which they apply. This tends to promote the technical reasonableness of the requirements and the measurements and assessments, as well as the workability of the procedures. This policy also tends to create the atmosphere for line organization ownership of the requirements, measurements and assessments, and procedures. Although the requirements may emanate from the law, through a corporate-level environmental department, this policy enables the line organization to have an important voice in how the requirements are ultimately conveyed and worded. Also, the line organization and the environmental organization have an equal voice in the establishment of the methods of measurement and assessment. With regard to the procedures for attainment, the line organization has the final voice.

This policy also emphasizes the need for procedures to focus on preventing noncompliance, recognizing that the effects of some noncompliances are irreversible or reversible at a cost that exceeds the cost of prevention in the first place. Of course, in those rare cases for which the cost of absolute prevention significantly exceeds the cost otherwise (taking into account all costs, including political costs) it may be well to settle for less than absolute prevention. In such cases, prevention still should be the objective of the procedures but it need not be absolute prevention, so as to enable compatibility with a policy to be covered later dealing with the benefit-risk relationship of any environmental action.

This policy also calls for the implementation of a system by which to recognize the need for changes to the environmental requirements and by which to control such changes. The purpose of a change control system is to assure that the requirements and procedural information being conveyed is timely and correct in light of changes in the law; in the design of facilities, equipments, and processes; and in environmental control technology, as a whole.

A policy that goes hand-in-hand with the preceding ones is that

Objective evidence of compliance with environmental requirements shall be maintained.

This means that the results of the quantitative measurements or qualitative assessment are to be documented in a way that permits a demonstration of compliance. If the quantitative measurements are made with automated equipment, the chart recordings or the measurement equipment logs and calibration records are to be used to demonstrate that the equipment was operating accurately and, in turn, to demonstrate compliance. Such demonstrations are necessary because they increase confidence in compliance, permit independent assessment, and provide legal protection.

Another typical policy is that

Activities relating to environmental compliance shall be audited by an organization that is independent of the line organization.

The most salient point of this policy is the establishment of the audit function. While, on the one hand, the audit function is certainly a significant part of an overall program by which to manage for environmental control, sometimes too much dependence is placed on audit alone. An audit alone does not provide the policies, requirements, and procedures necessary for economical environmental control. Nor does it provide the in-line environmental measurements necessary to be performed in real or in reasonable time. An audit alone does not provide the impetus to the prevention of environmental noncompliance. An audit is a valuable element of a total environmental control program—an element geared toward identifying inadequacies in the overall management system and in identifying noncompliances with the system.

ENVIRONMENTAL POLICIES

Another salient point of this policy is that the audit function shall be performed by an organization that is independent of the line organizations in order to contribute to the objectivity of the audits.

The following policy almost goes without saying. It is that

Appropriate officers shall be made aware of any environmental condition which may significantly affect the company.

This policy stems from corporate, regulatory, and community ultrasensitivity to environmental matters. It applies not only to that which is done but also to that which is not done. For example, an inadvertent spill, on the one hand, and a decision to not utilize a spill prevention technique, on the other hand, may equally significantly affect the company in the long run.

This policy applies not only to accident or incident situations. The environmental department must identify to the appropriate officer any unresolved significant weakness in the total environmental program.

As somewhat of a corollary to the preceding policy, another typical policy engendered by the sensitivity to environmental affairs is that

Appropriate officers shall be made aware of any facility or equipment modification or operating procedure change which can significantly influence the company's ability to attain compliance with environmental requirements. Officers shall be made aware of the modification or change in advance of the company's commitment to it.

This policy works in both directions, being applicable to modifications or changes that either enhance or detract from the ability to achieve compliance. On the one hand, the company should be assured that for any environmentally beneficial modification or change, the cost of the benefit is reasonable. The company wants to be assured, also, that the maximum regulatory and community relations benefits are being derived from such a modification or change. On the other hand, for a modification or change which may work in the opposite direction, the company must be assured that full risk is assessed, that the justification for taking such a risk is substantiable, and that the company has prepared itself for the possibility of the adverse impact actually taking place.

Still another related policy is that

Environmental compliance alternatives shall be addressed on the basis of benefit-risk analysis. Agreement on the alternative selected shall be obtained from the responsible officers in charge of the engineering department, if design is involved, the operating department, and the environmental department.

The purpose of this policy is to help to assure, on the one hand, that the least costly method of complying is not automatically chosen simply because it is the least costly while, on the other hand, some high cost method is not unjustifiably chosen simply because of external regulatory or community pressures, or for that matter, the zealousness of the environmental department. For each alternative, the potential cost of the risks must be weighed against the potential cost avoidance of the benefits. (To me, the term *benefit–risk analysis* makes a lot more sense than the term *cost–benefit analysis*. The objective of the analysis is to determine what cost may be incurred by taking certain risks and to determine what cost may be incurred by not taking those risks. By comparison, the lesser of the two costs may be selected. It is really an analysis of the benefits and risks, using cost as the common measurement denominator.)

The other salient point of this policy is the requirement for unanimity of agreement with regard to the selection of any alternative which is intended to satisfy an environmental requirement. This policy recognizes that the interests of a variety of stakeholders must be satisfied and that the environmental department is in the best position to recognize the interests of the regulatory and community stakeholders. This is not to say that the environmental department is to serve those interests. The environmental department, like every other department of the company, is to serve the company's interests. Ideally, the company's interests are best served by appropriately recognizing and dealing with the interests of the regulators and the community, as well as the needs of the company.

One of the most significant policies bearing upon the relationship between the regulator and the company is that

> **The environmental department shall attempt to achieve before-the-fact agreement with the regulatory agency as to the scope, method, decision criteria, and implementation schedule of any study, the results of which are intended to be used by the company to arrive at a company position relative to the law.**

There have been too many cases in which a company has performed an environmental scientific study only to find that the study scope or method did not satisfy the regulator, possibly because the scope was too narrow or the method insufficiently conservative or objective in the opinion of the regulator. Often, even when the scope and the method of a study yield results with which both parties can agree, the parties apply different value systems to the results. In other words, the regulator may perceive the need for action at a lower threshold level than was anticipated by the company. Both parties may agree on the environmental effect of a certain company operation but the regulator may be unwilling to accept the effect whereas it may be acceptable to the company. Scientific method requires a predetermination of acceptance criteria such as to prevent the acceptance criteria from being unduly adjusted when the results become available and do not support one or the other party's predisposition. The acceptance criteria may be swayed by the study results and the leanings or biases of those using the results.

Undoubtedly, one may conceive a variety of additional policies or different wording of these policies. That would be fine. The intent here is merely to convey very broad concepts which may have to be adjusted based on the circumstances specific to each company.

TYPICAL ENVIRONMENTAL DEPARTMENT RESPONSIBILITIES AND AUTHORITIES

In the management systems hierarchy, organizational charters exist to define the responsibilities and authorities of each organization so as to enable the company to abide by the policies. The following are some responsibilities and authorities which should be typical for an environmental department.

Possibly the main role of the environmental department should be to

Contribute to company environmental policy and establish environmental standards and requirements.

This is a functional role as described earlier in that the standards and requirements must apply laterally, across-the-board, to all other organizations of the company. The environmental department should not be limited merely to setting standards and requirements for its own organization.

On the other hand, if there is to be willful compliance with these standards and requirements, they cannot be established in a vacuum. They must be established with the participation and equal partnership of the other functional and line organizations which are impacted by the standards and requirements. Teamwork is not achieved when one organization acquiesces to the needs of another organization at the expense of its ability to achieve its own objectives or to achieve its objectives to the same degree as are achieved by other organizations.

An equally important role of the environmental department should be to

Assure the establishment of procedures by which to comply with the policy, standards, and requirements and by which to measure and assess whether or not compliance has been attained.

This, too, is largely a functional role in that, for the most part, the procedures are of an interdepartmental nature, crossing departmental lines. For the most part, the procedures must be implemented by other than the environmental department.

In assuring the establishment of procedures by which to comply with the standards and requirements, the environmental department should concern itself with the completeness, logic, specificity, and clarity of the procedures. The environmental department should be concerned with the communications quality of the procedures. Of course, for each procedural action, there must be an assignment of a responsible individual and he or she must be capable through education, experience, and training in the specific contents of the procedure.

In addition, for measurement and assessment procedures, the environmental department should concern itself mostly with the effectiveness of each measurement/assessment—i.e., whether or not the right parameter is being measured/assessed at the right time; the consistency of each measurement/assessment—i.e., whether or not the measurement/assessment technique is described with sufficient specificity to yield data that are related to one another over time and data in which the line and functional organizations can have confidence; and the economy of each measurement/assessment—i.e., whether or not it is performed in the least costly fashion, itself, and has the least adverse cost impact on the line organization.

In many cases, it is not economically viable for the environmental department to staff itself with the kind of expertise necessary to determine, at the outset, whether or not an operating procedure will yield compliance with environmental standards and requirements. Therefore, the environmental department must rely on the line operating organization for the application of this expertise to the procedural development and review process. However, the environmental department should have the expertise to determine, from the outset, the adequacy of environmental measurement and assessment procedures. In the absence of such expertise in the environmental department, there would be no independence of the measurement and assessment process and its objectivity could be questioned.

There are times when the environmental department should acquire outside consultation, not only to fill a technical capability gap, but also to add perceived objectivity and credibility to the benefit of ensuing negotiations with the regulator or community.

Another major environmental department role should be to

Acquire environmental data and determine the need for corrective action.

This is mostly a line role. Although in many cases environmental measurements may be made by other than the environmental department, the procedures undoubtedly require the environmental department's timely review of the measurement and assessment data. In turn, the environmental department, through its review of the data, should provide the ultimate determination of the appropriateness of continuing the process. The hope is that the operating organization, through its own timely review of the data, will make its own stop work decisions, when necessary.

Obviously, when it is decided that operations cannot continue, it is also decided de facto that there is a need for operational corrective action. The environmental department must also decide on the need for the correction of the underlying or root cause of the operational problem instead of merely the correction of the operational noncompliance at hand.

Related to its responsibility to determine the need for corrective action, the environmental department also should be responsible to

Assure the attainment of environmental corrective action.

This too is largely a line role. The suitability of returning to operational status should be judged independently by the environmental department. In addition, the determination as to whether or not the root cause of an environmental problem has been identified and eliminated should be made by the environmental department—presumably the department's objectivity being its strong suit in addition to its technical capability.

Another responsibility and authority of the environmental department should be to

Acquire environmental permits, licenses, and certifications.

Although the input from many other departments is usually required in order to process applications for permits, licenses, and certifications, the environmental department is in the best position to manage the process because of its overall and specific knowledge of the regulatory requirements and because of its day-to-day interface with the regulatory agencies that issue the permits, licenses, and certifications. Sometimes, at the local governmental level, it is best for the operations organization to assume this line function because it, and not the environmental department, interfaces on a day-to-day basis with the local regulator.

Another responsibility of the department should be to

Provide the company's position on environmental science and compliance aspects of proposed legislation, rules, and regulations.

This is a line role. However, the environmental department should not have exclusive responsibility and authority in this regard. Rather the

ENVIRONMENTAL POLICIES

department should be responsible for pulling together the comments of the various departments and resolving any conflicting comments. Again, the department should be managing the process.

Finally, it should be the environmental department's role to

Represent the company before regulatory and industry groups on environmental science and compliance matters.

Again, this is not intended to be an exclusive responsibility and authority. For example, in the event of a legal action, the legal department certainly is expected to take the lead in developing legal strategy but the environmental department should take the lead in helping the company to define its technical strategy and in providing the testimony in support of that strategy.

This is not intended to be a complete description of the responsibilities and authorities of an environmental department but rather a description of the major ones. If the environmental department does not have at least these roles, the level of the company's maturity with regard to environmental affairs may be questionable.

3

Environmental Requirements Documents

INTRODUCTION

On a facility-by-facility basis, in addition to annual environmental performance improvement goals, the environmental management system should require the preparation and implementation of environmental requirements documents containing at least two elements: first, the element that specifies the environmental performance levels required to be achieved and second, the element that specifies the measurements and assessments to be made by which to determine whether or not the required performance levels have been achieved. This chapter provides a discussion of the information that should be included in the environmental requirements documents. This chapter also provides a discussion of the attributes of a requirements document needed to make it a good communications medium.

FACTORS TO BE ADDRESSED IN ENVIRONMENTAL REQUIREMENTS DOCUMENTS

The first question to be answered is whether to establish the requirements on a company-wide or a facility-by-facility basis. Even when the operations at one facility are identical to the operations at another facility, which is quite unlikely, there still may exist differences in the requirements, based on the vintage of the facility or on differences in state and local law and levels of community sensitivity. In determining and documenting the requirements on a facility-by-facility basis, there is greater assurance of a real understanding of the requirements. Furthermore, on this basis, the environmental department can better convey the perception of its willingness to accommodate facility differences. The perception is as important as the reality. Another way of saying the same thing is that documents established on a facility-by-facility basis also provide for the facility to have greater self-determination with regard to its environmentally related operations. With greater self-determination comes greater ownership—ownership of the requirements and ownership of the document in which the requirements are stipulated. Facility-by-facility based documents better enable the operations personnel to "buy into" the requirements and to establish an environmental performance "contract" with the environmental department.

The environmental requirements document for the facility should specify each performance requirement applicable to the facility. The performance requirement should be stated in terms that are understandable to the operating personnel. The performance requirement should not merely be a quotation from its lawful source. The requirement should be an individual parameter, which is individually applicable to the facility, individually assignable to a person or a first level organization, and individually measurable or assessable.

The requirement must be attainable within the constraints imposed by the facility, the equipment and the personnel capabilities. If the requirement is unattainable, either the requirement must be relaxed, if permissible, or the capability of the facility, equipment, or personnel must be improved, or both, before the facility should be allowed to operate. In the event that the facility operates consistently in violation

ENVIRONMENTAL REQUIREMENTS DOCUMENTS

of a requirement, in the long run, disrespect for environmental requirements as a whole will accrue. In these times of regulatory sensitivity and increasing power of the authorities to hold lower level personnel accountable, it is doubtful that any operations supervisor would want to assume the legal responsibility for consistently and knowingly operating in a manner that yields noncompliance with environmental law.

For each environmental performance requirement, the specific measurement or assessment to be used to determine the state of compliance with that requirement should also be specified. A performance requirement is meaningless without a corresponding performance measurement or, in the absence of a measurement, at least a qualitative assessment. The method for making the measurement or assessment should be specified in the requirements document directly or by reference to another procedure. The place and time at which the measurement or assessment is to be made should be specified. The requirements document should specify the person or organization responsible for making the measurement or assessment.

For each measurement or assessment, the minimum information to be recorded should be specified along with the format for recording the information, the timing of the recording, and although it may be obvious, the person or organization responsible for the recording (normally the person or organization responsible for making the measurement/assessment).

For each measurement or assessment, the requirements document should specify the person or organization (in addition to the environmental department) responsible for reviewing the results of the measurement or assessment and for making a decision as to the action to be taken in light of those results. The action steps may be specified in the requirements document or may be specified in a separate procedure referenced in the requirements document.

Undoubtedly, it is preferred that environmental compliance be attained by means of the automated design capabilities of the facility and equipment—design capability to automatically take measurements, feedback information, provide alarms, and take corrective action, including automatically shutting down operation when no other corrective action is readily available. The greater the need for operator intervention in the attainment of operating corrective action and the greater the potential adverse impact of noncompliance, the greater the need for automated controls in real time.

As discussed in Chapter 1 and depicted in Fig. 1, the effectiveness of a total environmental control program is, to begin with, dependent in large part on the adequacy of the design of the facility and equipment systems as well as the management systems.

The requirements document should be approved by each organization whose functional or line responsibility is impacted by the document. The requirements document should specify operating requirements or limitations by which to achieve compliance with the environmental requirements. Therefore, as a minimum, the line operating department and the environmental department should approve the document.

This alone will not assure the operating department's "buy in." This alone will not assure the existence of a real contract between the parties. From the outset, the requirements document has to be developed jointly if there is to be real acceptance of it by the parties. The operating department must come to understand the bases for the environmental constraints and the environmental department must come to understand the impact of those constraints upon production. Without this understanding at the organizations' lowest implementation levels, it will be harder to gain consistent compliance with those requirements. Simply putting together a set of operational requirements and a set of environmental requirements, each of which have been developed independently, will not get the job done effectively.

The requirements document should be subject to a change control process. As a minimum, the change control process should enable the identification of any necessary change to a requirement, establish the approval signatures that are prerequisite to making the change, the mechanical means by which to make the change, the means by which to assure that the change has been implemented, and the means by which unauthorized changes are prevented from being implemented.

EXAMPLES OF REQUIREMENTS DOCUMENTS

Appendixes 3.1, 3.2, and 3.3 provide examples of requirements documents for a coal-fired power plant.

ENVIRONMENTAL REQUIREMENTS DOCUMENTS

Appendix 3.1 Requirements Document—Example 1

A. *Requirement:* Groundwater quality measurements and reports as specified below
   1. Basis: Solid Waste Management Act 641 and promulgated rules (Rules 299.4305, 4306, and 4315). Schedule of Compliance dated March 26, 1980 between the MDNR and the company.

B. *Measurements*
   1. What: The following sixteen parameters shall be measured in each groundwater sample: arsenic, barium, bicarbonate, calcium, carbonate, chloride, chromium, iron, lead, magnesium, pH, selenium, sodium, specific conductance, sulfate, and total dissolved solids.
   2. When: The sampling shall be made by December 5th of each year.
   3. Where: Three monitoring wells, numbered 79MW-3, 82MW-1, and 82MW-2, around the perimeter of Ash Pond 6 shall be sampled.
   4. How: Each groundwater monitoring sample shall be collected in compliance with the Water Quality Monitoring Program (WQMP) for the J. R. Whiting Plant.
   5. Who: The PIS shall coordinate onsite activities and assure that each required sample is collected and analyzed by an approved testing laboratory.

C. *Reports*
   1. What: The following shall be provided:
      a. Written notification indicating each monitoring station sampled, the date on which the sample was taken, and any potential problems.
      b. A complete report of results.
   2. When: The results shall be reported as scheduled below:
      a. Written notification shall be provided within three days after each sample is taken.
      b. A complete written report of results shall be furnished no later than December 31st.
   3. How: The written report shall be made in compliance with the WQMP.

Appendix 3.1 *(continued)*

4. Who:	The PTS shall assure that the laboratory: a. Provides written notification per 2a, above, to the Aquatic and Solid Waste Section of the ED. b. Provides the report of results per 2b above, to the Aquatic and Solid Waste Section of the ED.

Appendix 3.2 Requirements Document—Example 2

A. *Requirement* The condenser cooling water discharge from Outfall 001 shall not exceed a monthly average concentration of 0.2 mg/l or a daily maximum concentration of 0.3 mg/l of total residual chlorine (TRC). Chlorine application time shall not exceed 160 min in any 24-hr period. Chlorine application time shall be reported.
   1. Basis: Permit, Page 2 of 12, Part IA1.
B. *Measurement*
   1. What/When: The TRC concentration shall be measured during each chlorine treatment.
   2. Where: The TRC concentration shall be measured from samples obtained at the headwall where the discharge occurs.
   3. How: Three grab samples equally spaced during the duration of the treatment shall be obtained and the TRC concentration shall be measured by the amperometric titration technique in accordance with EPA-approved analytical testing procedures (40 CFR 136.3).
   4. Who: The CO shall assure the TRC concentration measurements are correctly made and that the application time is properly determined.
C. *Written Reports*
   1. What: The TRC concentration measurements and chlorine application time shall be reported in the MOR form issued by MDNR. For nondischarging days, the MOR reporting space should be left blank.
   2. Who: The CO shall submit the MOR form to OSD for their transmittal of copies to the MDNR Water Quality Division, ED, and the Plant.
   3. When: The form shall be submitted to OSD not later than the fifth working day of each month. The copies shall be submitted to MDNR, ED, and the Plant by the seventh working day of each month.

Appendix 3.3 Requirements Document—Example 3

A. *Requirement* The daily average of the stack sulfur dioxide emission rate shall not exceed 1.2 lb of sulfur dioxide per million Btu of heat input to the boiler.
   1. Basis: EPA Reg 40 CFR 60.8(a), 40 CFR 60.43(a) (2), and 40 CFR 60.45. MAPCC Reg. 336.1912 and 336.2001.

B. *Measurements—Continuous Emission Monitoring*
   1. What: Sulfur dioxide emission rate shall be measured.
   2. When: The measurement shall be continuous.
   3. Where: The measurement of the average emission rate shall be made in accordance with Appendix A and shall be recorded in the environmental log each hour.
   4. How: By monitoring the sulfur dioxide emission rates recorded at the plant environmental log.
   5. Who: The SWS shall review the log hourly to assure that the log entries are made.

C. *Event Report*
   1. What: An event report shall be made to cover the following:
      a. Any occurrence of sulfur dioxide emissions in excess of the requirement specified in A, above.
      b. The duration, magnitude, and cause of the excess emission.
      c. The corrective action taken to reduce emissions to compliance levels.
      d. The corrective action taken to prevent recurrence.
   2. When: The report shall be made no later than the next business day following the day on which the exceedance occurred.
   3. Who: The SWS shall notify ED. (ED shall determine if it is necessary to notify MDNR.)
   4. How: Orally. If during the oral notification of the event, ED determines that a written report is required, it shall be prepared by the CO and submitted to ED within five working days of the occurrence. The written report shall discuss the items specified in C.1.a-d, above. (ED shall transmit the report to MDNR.)

D. *Monthly Report*
   1. What: A monthly report shall be made to cover the following:

ENVIRONMENTAL REQUIREMENTS DOCUMENTS

Appendix 3.3 *(continued)*

        a. Any sulfur dioxide emission rate in excess of the requirement specified in A, above; the duration, magnitude and cause of the excess emission; the corrective action taken to reduce emissions to compliance levels; and the corrective action taken to prevent recurrence.

        b. The date and magnitude of each 3-hr block average sulfur dioxide emission rate in excess of 1.2 lbs per million Btu.

        c. The date and time of each period during which the sulfur dioxide monitoring system was inoperative in excess of two consecutive hours or four hours total in one day, the factors that precluded monitor operations, and the nature of the repairs or adjustments made.

    2. What: The report shall be on a monthly basis, by the seventh day after the end of each month.

    3. How: The report shall be written per the format given in Appendix A.

    4. Who: The CO shall prepare the report and submit it to the ED by the eighth day after the end of each month. (ED shall transmit the report to MDNR.)

E. *Measurements—Performance Testing*
    1. What: Sulfur dioxide emission rate.
    2. When: Sulfur dioxide emission tests shall be made within 60 days following the ED's receipt of written notification from the MAPCC that such tests are required for compliance assessment.
    3. Where: The tests shall be made at test port locations approved by the MAPCC as specified by ED.
    4. How: The tests shall be made according to the performance test criteria specified in Part 10, Appendix A of the MAPCC Rules.
    5. Who: The tests shall be supervised by OSD.

F. *Performance Test Report*
    1. What: The report of the results of the sulfur dioxide emission measurements specified in E, above, shall include:

Appendix 3.3 *(continued)*

	a. Sulfur dioxide emission rate.
	b. Flue gas conditions (temperature, volume, moisture, excess air).
	c. Boiler conditions (steaming rate).
	d. Fuel characteristics.
	e. Description of test port locations and testing procedures.
2. When:	The test report shall be submitted within 45 days following test completion.
3. Who:	The OSD shall prepare the report and submit it to ED. (ED shall transmit the report to MDNR.)

4

Environmental Considerations in Facility Site Selection

INTRODUCTION

This chapter describes the factors to be considered in establishing a management systems procedure for the selection of a site for a new facility. The procedural approach recommended in this chapter is aimed at minimizing the environmental department and company expenditures while at the same time assuring that the sites that ultimately become candidates for selection are environmentally qualified. Site environmental qualification is obtained through a process of progressively more comprehensive site data collection with the data being compared to progressively more stringent sets of exclusionary environmental criteria. In progressing from one phase of the process to the next, the scope and quantity of data collected for a given site is increased. At the same time, however, the number of sites under consideration is reduced.

The intent of the procedure should not be to yield the best site from an environmental viewpoint but rather to yield a reasonable number of alternative sites, each of which is wholly adequate from an environmental viewpoint. The flexibility provided by a number of alternative, candidate sites enables the company to choose the site that has the best overall advantage. Obviously, such factors as site construction cost, labor availability, taxes, proximity to sources of supply and markets, political climate and local attitudes, and a host of others apply to the site selection process. The intent here is to assure that the site to be chosen will meet the minimal environmental requirements and, beyond that, to assure that the environmental features of the site will be appropriately considered, along with all of the other factors to be considered, in the ultimate site selection.

In the event of significant economic advantage to the selection of a site which does not meet the minimal environmental criteria, the process should assure the application of technology to eliminate the environmental problem and to render the facility in full compliance with the environmental requirements before operations are begun. This is fundamental. For example, if a home building site does not pass the "perk" test to allow the use of a standard septic system, the site still can be used by designing and installing a special septic system to accommodate the poor drainage qualities of the site and to render the homesite in full compliance with the sanitation and health laws. In other words, design, construction, or process control commitments may be made to overcome an environmental problem with the selection of a given site. There must be assurance, however, that the commitments will adequately address the problem and that, once made, the commitments are adhered to.

For want of a better place at which to address the subject, this chapter concludes with some thoughts on how to protect the company from environmental noncompliance when the company leases its land to another party.

PHASE I: IDENTIFYING POTENTIAL SITES

The steps in the site selection process are summarized in Fig. 5. The first step in the process should be to define *candidate regions* of interest and within each region to further define, in general, the characteristics that may affect the environmental suitability of the region or of any of its parts. A good way to do this is to identify

FACILITY SITE SELECTION

Phase I

A. Identify the *candidate regions* of interest.
B. Identify the *potential subregions,* which are generally homogeneous.
C. Define the environmental suitability characteristics that may affect the suitability of each subregion.
D. Scrub the subregions that have unsuitable characteristics. Retain the *candidate subregions*.
E. Identify sites within the subregions.
F. Establish the environmental exclusionary criteria for *potential sites*.
G. Scrub the sites that do not meet the criteria. Retain the potential sites.

Phase II

A. Develop environmental data bases (low cost data) for the potential sites.
B. Compare the site data bases with the suitability characteristics (I.C, above and exclusionary criteria (I.F, above).
C. Scrub the potential sites that do not meet the suitability characteristics and exclusionary criteria. Retain the potential sites that do.
D. Establish more specific environmental constraint criteria.
E. Establish the ratings that are to be applied to the site condition relative to each constraint criterion (II.D, above).
F. Compare the data (II.A, above) for the sites with the more specific constraint criteria (II.D, above) and apply the ratings (II.E, above).
G. Scrub the problem sites. Retain the *Candidate Sites*.

Phase III

A. Identify candidate sites with environmental data voids.
B. Scrub the sites for which there is a high cost to collect the missing data.
C. Collect the missing data for each candidate site.
D. Compare the new, additional data, with the specific constraint criteria (II.D, above) and apply the ratings (II.E, above).
E. Scrub the candidate sites that are unpermittable/unlicensable. Retain the *Potentially Permittable/Licensable Sites*.

Figure 5 Summary of site-selection process.

potential subregions within the region that are generally homogeneous with respect to such things as population patterns, water sheds, coastal zones, topography, land use, and other similar attributes. The idea, at this point, is that a subregion may contain potential sites because of the subregion's general homogeneous character. Certain subregions will be excluded at this point because of their general character, other subregions can be classified as *candidate subregions*.

Having identified candidate subregions, the next step is to identify potential sites within the subregions. This can be done by establishing environmental exclusionary criteria relating to such things as the availability of a cooling water source, the availability of sufficient assimilative capacity to absorb waste without deleterious effects, the ease of housing and transporting waste, the presence of geologic anomalies, and the presence of dedicated lands. When these exclusionary criteria are applied to the candidate subregions, certain subregions or portions of subregions will fail to meet these criteria and may have to be excluded from further consideration. The unexcluded areas may be viewed as *potential sites*.

Notice that up to this point in the process not much company resource has been expended for environmental considerations. The information necessary to define the candidate subregions is generally readily available. The definition of the environmental exclusionary criteria and the application of these criteria to arrive at potential sites is also not too difficult.

PHASE II: IDENTIFYING CANDIDATE SITES

In this phase, the first step is to develop the environmental data bases for the potential sites. To this point, the data base should focus on the types of site characteristics which will be critical to obtaining the construction permit or operating license. The data need not be comprehensive in scope. Also, for any environmental characteristic, the data need not be detailed at this point. Additionally, the data need not be based on intensive scientific study but may be based merely on field observations.

Now the data base for each potential site should be compared to the existing suitability and exclusionary criteria for the subregions and

FACILITY SITE SELECTION 45

sites. Potential sites that do not meet the constraints may be excluded from further consideration.

Next more specific environmental constraint criteria should be established. These criteria may relate to:

- The prevention of significant deterioration of air quality
- The attainment of air quality for critical pollutants
- The potential impact of water withdrawal on the aquatic resource
- The prevention of deterioration of surface water quality
- The river seasonal flow rates for assimilative capacity
- The suitability for solid waste disposal
- The prevention of deterioration of ground water quality
- The encroachment upon wetlands
- The encroachment upon flood plains
- The local presence of threatened or endangered species
- The identification of lands and waters recognized to be of high ecological or wildlife value
- The proximity to dedicated lands and waters
- The encroachment upon state designated lands or upon lands of restricted use
- The potential impact on areas of high scenic quality
- The suitability of electric transmissions options
- The potential impact on social and economic values

These environmental constraint criteria must be specific except in the case where that which is unacceptable is specified rather than that which is acceptable, as is customary.

Also, at this point in the process, it is necessary to establish the ratings that are to be applied to the site conditions relative to the environmental constraint criteria. The following, for example, are six rating levels that may be used:

1. There is "no constraint" upon the potential site because the criterion is not applicable to the site.
2. There is a "slight constraint" on the use of the potential site. This means that the potential site does not satisfy the exclusionary criterion but that the negative impacts can be avoided through prudent planning. This rating also means that

no delay or uncertainty is expected to result in the permitting or licensing process.
3. There is "moderate constraint" upon the use of the potential site. This means that the negative impacts of the site's noncomformance to the exclusionary criterion can be avoided through some special design or operating requirements or by incurring some special mitigative expense. For this rating, also, no delays or uncertainties are expected.
4. There is a "serious constraint" upon the use of the site. This means that a considerable modification of design or operating requirements or a considerable mitigative expense is required to overcome the problem. In this case, permitting or licensing delays and uncertainties could result.
5. There is a "severe constraint" upon the use of the potential site for a given exclusionary criterion. This rating means that in addition to the considerable expense likely to result from design or operating modifications and mitigations, legal proceedings are likely to occur and, as such, there is a possibility that the potential site could be rendered unsuitable for permitting or licensing.
6. There is an "exclusionary constraint" upon the site. In other words, the site should be dropped from further consideration. This might be attributable to unacceptable levels of expense or known legal entanglements which undoubtedly will cause extensive delays, if not failure, to acquire the permit or license.

Now the analysts are in a position to compare the environmental data for each potential site with the exclusionary criteria. In each case for which the criterion is not satisfied by the data, the analyst is further able to rate the severity level of the problem. If there is an abundance of potential sites, the line of demarcation may be drawn even tighter than above the sixth rating, possibly above the fifth or even the fourth rating. From this process emerge the *candidate sites*.

Even to this point, minimal company resources have been expended for the acquisition of environmental data. The data which should be used to this point should be readily available through low cost field

FACILITY SITE SELECTION

observations or surveys, governmental publications, and other low-cost sources such as from companies already located in the potential sites which are willing to share their data at little cost.

The administrative process being described here is on an exception basis—in other words, on the basis of progressively excluding sites for which environmental data indicates a site's unsuitability.

PHASE III: IDENTIFYING POTENTIALLY PERMITTABLE AND LICENSABLE SITES

At the beginning of this phase there will most likely be candidate sites for which environmental data is lacking. This data must be acquired in order to determine the site's permittability or licensability. Obviously, if there are many candidate sites at this juncture, any site for which major data voids exist and for which data collection costs would be high, may be excluded from further consideration if there is no outstanding economic benefit to that site. Why spend a large amount of money for data collection to determine the potential permittability/licensability of a site when there are many other such sites to fill the bill at far less cost? The environmental department within a company, whose business is not scientific research, should not engage in scientific data collection simply for the sake of science. On the other hand, this process should not be performed in an environmental vacuum. If a candidate site has unique economic benefits which more than compensate for the cost of the additional data acquisition, the site should be retained as a candidate.

After identifying the additional environmental data needed to supplement the existing data base for a given candidate site, the next step is to specify the data collection requirements. Specifically, what kind of data should be collected? In what quantity is it to be collected? From what points should it be collected? What method is to be used for its collection? How accurate should it be? How is the raw data to be processed? How is it to be analyzed? What means are to be used to assure the accuracy of the data collection and data processing? Are the data collection, processing, and analytical techniques universally acceptable? What are to be the qualifications of those who collect, process, and analyze the data? These are typical of some of the questions that must be answered in the specification. These questions apply to each type of data element to be collected.

An important point, which will be stressed later in the chapter dealing with the management of environmental study projects, is that prior to the collection of environmental data there should be agreement between the company and the regulatory agency as to the scope of the data to be collected; the method of collection, processing, and analysis; the level of accuracy; and the methods for assuring the attainment of this accuracy. Assuming that this data will be used to demonstrate the ability of the site to meet regulatory environmental requirements, it makes little sense to risk the possibility of the data being unacceptable to the regulatory agency because of its insufficiency, because the accuracy of the data or the degree of compliance with the data collection and data processing methods is questionable, because the method by which the data was collected was unacceptable, or because the data analysis technique is unconservative. A great deal of money and, more importantly, time can be spent collecting and processing additional data to satisfy regulatory requirements and concerns. Remember, the burden of proof of acceptability is on the permit or license applicant and if the regulator simply does not "feel comfortable" with the situation, that will be enough to cause delay in the application's progress.

Much of the environmental data collection is seasonally constrained. The consequential cost of the delay while collecting this data can be many times the cost of the data collection itself. It is essential, therefore, to bring the regulatory agency into the process as soon as possible and to get agreement on these aspects of data collection and analysis activities before these activities are begun.

Having collected the additional data required for each candidate site, the next step is to compare this data with the environmental constraint criteria and the environmental ratings discussed earlier. This step is essentially identical to the corresponding step in the previous phase.

Some candidate sites may be clearly unpermittable or unlicensable and may be dropped from further consideration. Some candidate sites may be of marginal permittability or licensability and their further

FACILITY SITE SELECTION 49

consideration is dependent upon the comparison of the cost to make them permittable/licensable versus the economic benefits to be derived from their use. In any event, a line of demarcation must be drawn above which the sites may be classified as potentially permittable/licensable.

It is important to note that this process of identifying potentially permittable/licensable sites cannot be accomplished without the help of design engineering, operational, and legal personnel who contribute information beyond the scope of information usually available to environmental department personnel. Who but the design engineer can best identify facility modifications and the cost of such modifications to overcome an environmental problem? Similarly, who but the operations expert can best identify techniques and their cost by which to overcome an environmental problem? Who but an attorney can best determine the potential legal constraints of permitting or licensing a site? At this juncture, the potentiality permittable/licensable sites can be ranked both from an environmental viewpoint and from an overall viewpoint.

Remember, the major objective of the environmental department is not to cause the selection of the site which is the best environmentally. Rather, it is to contribute to the selection of the site which is the best from an overall company viewpoint while, at the same time, assuring that the site either meets all environmental requirements or can be made reasonably to meet all environmental requirements. In the latter case, the environmental department should secure the commitments up front. Before pursuing the permit/license, there should be an understanding on the part of all departments as to what is needed to comply with the law and company commitments.

The environmental department should set the stage for a good working relationship with the regulatory agency. The things that are done in the process of identifying potentially permittable/licensable sites will carry through the construction and operational phases of the project.

Finally, the environmental department should collect the environmental data or have it collected as discreetly as possible. This is an important objective recognizing that, after site selection, the company may have to purchase the site or purchase an option on the site.

It is important to involve the public. No matter how well conceived the data collection and site selection process may have been, there is no way to fully anticipate public reaction to a project. A positive interaction with the public at this point will pay dividends. The cost of satisfying public demands at this point can be far less than the cost of public controversy.

FURTHER CONSIDERATIONS

Throughout the process it is important to document the rationale for the exclusion of each site from further consideration. There will be internal and external pressures for the retention of many sites that must be excluded for environmental reasons. The justifications for their exclusion must be logical and complete.

The greatest external pressure, however, will most likely come from the state environmental regulatory agency which may have a somewhat different agenda. The state's environmental professionals may bring pressure to bear upon the selection of the site which is best from strictly an environmental point of view—not the best from an overall company viewpoint. These divergent objectives need to be recognized and addressed as early as possible in the process. If this difference is not resolved up front, any other agreements relating to data collection, as discussed earlier, may be to no avail in maintaining a good working relationship with the state environmental agency. Often, this kind of resolution requires help at the policy making levels of both the company and the state. It is important for the environmental department to recognize the diverging objectives and to recognize the need for help as well.

Throughout the process, one major objective is to conserve company resources by acquiring costly environmental data only for those sites which are truly candidate sites and by limiting the number of candidate sites to some reasonable number. Another major objective is to maintain a good working relationship with the regulator and the public. By far the most important objective is to protect the company and the community by assuring that the selected site meets the requirements of the law and the company commitments or that the action necessary to comply with the law and company commitments are accepted and implemented as part of the approved project plan.

LEASING COMPANY LAND

A company may find itself in environmental trouble if it leases its land and the lessee does not comply with environmental law or commitments made by the company relative to the use of that land. Therefore, the management system should provide protective measures, the most basic of which is to have a standard protective clause written into each lease. Basically by means of this clause the lessee should be precluded from disposing of any waste without the company's consent. The leasee should also be precluded from generating or storing any material which is hazardous, may become hazardous, or is controlled by an environmental regulatory agency. The clause should also require the lessee to grant access to personnel and representatives of the company to facilitate their performance of environmental inspections by which to assure compliance with the requirements of this clause. An example of this clause is given in Appendix 4.1.

The management system should also require that the environmental department be provided with timely notification of any such lease so that the department may schedule the aforementioned environmental inspections.

If the prospective lessee takes exception to this clause, the management system should allow the company to consummate the lease under the following additional contractual conditions. In this case, as a prerequisite to the contract, the lessee should be required to submit a written description of the environmental control program by which the lessee will assure compliance with environmental law and company commitments, and to obtain the company's acceptance of this program. Of course, the environmental department should take the lead within the company in reviewing and accepting or rejecting the lessee's program.

A company's failure to exercise these controls may result in its being subject to severe public relations and economic penalties. The company owns the land, not the lessee. Therefore, in the eyes of the public and the regulators, the company is responsible for the appropriate use and maintenance of the land. Usually the lessee is much smaller than the lessor company and the media and the regulators will seek retribution from the larger company. Sometimes the lessee goes out of business and the lessor company is left holding the environmental bag.

Appendix 4.1 Standard Clause for Land Leasing

It is expressly understood that the lessee shall at all times keep and maintain said premises in a clean and sanitary condition and shall comply with all laws of the United States and of this state, or of any regulatory body of the United States or of this state, or of any other governmental or governing body which may now or hereafter have jurisdiction over the subject matter which is now, or may hereafter, made effective while this lease remains in effect. Without limiting the generality of the foregoing, it is expressly agreed that the lessee shall not dispose or suffer to be disposed of any waste material, whatsoever, upon the premises without the prior written consent of the lessor, and shall not, without the prior written consent of the lessor, store, use or maintain, or suffer to be stored, used or maintained, upon said premises any material which is or may be or become hazardous to human health or the environment or the storage, treatment or disposal of which is regulated by any governmental authority. The granting or withholding of any consent of the lessor under the terms of this paragraph shall be within the sole discretion of the lessor and the lessee shall, when requested by the lessor, promptly give to the lessor any information required by the lessor concerning products, substances, processes used, stored, maintained or undertaken by the lessee or on its behalf or with its approval upon said premises. The lessee agrees to indemnify and save the lessor, its successors and assigns harmless from all loss and expense as a result of the failure of the lessee, his employees, agents, contractors and invites to comply with the terms of this clause.

Furthermore, it is expressly understood that the lessee shall grant to the lessor, his employees, agents, contractors and invites access to the land and to the lessee's facilities to enable the performance of environmental measurements, inspections, examinations, tests, surveillance and audits by which the lessor may determine the degree to which the lessee is compliant with this clause. Such access shall be granted also to any regulatory agency which has the proper authorization. Such access shall be granted at any time during normal business hours. In the event of an emergency or in the event that conditions are such as to possibly create an emergency or possibly create a noncompliance with this clause, such access shall be granted at any time.

5
Environmental Considerations During the Design Engineering Process

INTRODUCTION

This chapter describes the management system controls that should exist in the design process. These controls are intended to provide assurance that environmental factors are taken into account for design decisions. These controls should be no different from the controls necessary to assure that other significant factors are also taken into account—factors such as the constructability, inspectability, and testability of the facility and equipment design; the reliability, maintainability, and operability (especially with regard to human factors) of the design; the cost of constructing and operating the facility to the design (i.e., the cost effectiveness of the design); and the degree to which the performance requirements are satisfied by the design, including regulatory performance requirements among which certainly are environmental requirements.

CHAPTER 5
DESIGN REQUIREMENTS

The management system should include a procedure requiring environmental specialists to review the various documents which constitute the body of environmental law so as to identify the physical and functional requirements emanating from this law. The procedure should also require that the requirements be broken down into three major categories—first, those requirements that can be satisfied only by the inherent features of the design itself; second, those requirements that may be satisfied either by the inherent design features or by the implementation of certain operating procedures or constraints; and third, those requirements that can be satisfied only by operating constraints. The procedure should emphasize that the desired way of satisfying an environmental performance requirement is through the existence of an inherent capability in the design of the facility or its equipment—preferably an automated capability. Once established, such capabilities are essentially constant, except for possible reliability degradation or periods of off-line maintenance. Relying on operator actions or interventions as the means of meeting the requirement results in having to accept a greater risk of noncompliance, which is certainly not the most desirable situation.

The procedure should also require that the importance of each environmental requirement be determined. Criteria should be established by which to judge the importance of each environmental requirement. In general, the factors to be considered are the potential for any noncompliance—i.e., the probability of the noncompliance occurring; the ease with which the noncompliance may be detected and corrected; the degree to which the effects of the noncompliance may be reversible; the cost of reversing these effects; and the kinds of effects, such as real or perceived health hazards, political effects, or regulatory interface effects, to mention a few. To make these kinds of judgments, expertise is required in the environmental, legal, health, safety, and public affairs area, as well as in design engineering and operations. It is appropriate to classify environmental requirements in terms of their importance because such classifications constitute a significant input to the design process. A requirement receiving the highest classification may result in a commitment to a designed-in, inherent capability rather

DESIGN ENGINEERING PROCESS

than an operational constraint, or it may result in providing redundancy in the design as one example of the means by which to guard against equipment failure. If an environmental requirement is of the highest order of importance and if inherent design capability is not available for meeting this requirement, there may be a need for a designed-in capability by which to automatically shut down operations when the requirement is not being met or, at least, to provide in-line, real time measurement and feedback, with suitable alarms, to the operating personnel.

The procedure should also specify the types of inputs to be provided by each organization for the decision making process. To some extent the types of inputs needed will be predicated upon the techniques used in the decision making process. For example, if trade studies are to be used to make the decision between inherent design capability versus operating procedural constraints, and if a common denominator of such trade studies is to be cost either avoided or incurred, there needs to be procedural methods established by which to acquire and analyze this cost data.

There needs to be a system by which to review continuously for changes to requirements and commitments and by which to identify those changes which must be implemented immediately and those which must be implemented by a certain milepost.

Another systematic consideration is how to convey the environmental requirements (including changes to the requirements) and their classifications to the design engineering organization in a controlled and timely fashion. An important feature of controlling information is the ability to assure that changes to the information are made when they are warranted and authorized, authorized changes are made in a way by which they are recognized and understood, and that changes that are not authorized are not made. The need for communicating environmental requirements changes to the design engineering organization may occur frequently. The environmental requirements established at the outset of a project may change as a result of managerial review and feedback from environmental regulatory agencies. Over time, during the construction and operating processes, the regulatory requirements may change, as well as the commitments to the community. There will be design and operational changes which, themselves, will impose new environmental constraints that need to be addressed concurrently with the implementation of the design or operational changes.

DESIGN REVIEW

The management system should require the independent review of the design for any portion of the facility or for any equipment that can impact the company's compliance with environmental law and commitments.

The system should require two types of design review—the first being the review of the design requirements. The purpose of the requirements design review should be to provide an independent assessment as to whether or not the requirements to which the facility and equipment are being designed will satisfy higher tier internal and external requirements, such as company policy and environmental agency rules and regulations. The purpose also should be to assess the reasonableness of the classification of importance assigned to each requirement as discussed earlier. Requirements design review is very important because if the facility and equipment are designed to lesser requirements than necessary there may be a considerable adverse cost impact to the company. The facility operation may have to be delayed or derated and the facility/equipment may have to be modified or some other mitigating expense may have to be incurred.

As a minimum, the design requirements should be reviewed by environmental engineers, other than those who established the requirement or who supervised their establishment, and by environmental scientists and engineers from the environmental department.

The second type of design review that should be required is the review of the detailed design. The purpose of which should be to provide an independent assessment as to whether or not the detailed design will yield a facility and equipment which meets the established design requirements which, in turn, will enable compliance with environmental regulatory requirements. For example, is each measuring device, as designed, of sufficient accuracy relative to the requirement for which the measurement is to be made? Is the device sufficiently reliable? Is the environmental equipment protected from harsh environments that can accelerate its failure? If the equipment is going to operate in a harsh environment there is a need to demonstrate that it has

DESIGN ENGINEERING PROCESS

the capability of operating to the specified requirements for the specified operating lifetime, in the specified harsh environment. The means by which to make this demonstration must be established. This demonstration may be made by testing the equipment or by judging the equipment to be sufficiently similar to other equipment which has been tested with successful results, or by providing statistical data to show that the equipment has been used successfully in an identical or harsher environment for a long period of time. A chapter could be devoted entirely to the process of assessing an equipment's capability to perform adequately in a harsh environment. This is beyond the scope of this book. The interested reader should refer to any good book on product quality. Equipment environmental qualification is a common practice in the defense, space, and nuclear power industries.

Does the design provide the means by which to recognize that the device has failed? When the device has failed is there a redundant device or alternative means by which to make the required measurements?

There are literally dozens of design features that must be addressed in a review of the detailed design for environmental compliance. Therefore, it makes sense for the reviewers to use design review checklists as tools in helping to assure that the reviews are technically complete.

As a minimum, the participants in the detailed design review should include environmental engineers who did not participate in or supervise the detailed design, environmental scientists and engineers from the environmental department, representatives from the line organizations who have operating and maintenance expertise, and quality assurance engineers who have expertise in equipment reliability and maintainability, who have a useful history of facility and equipment problems, and who have to plan the facility and equipment inspection and tests.

The management systems procedure should require that the design review be supported by appropriate analyses, such as cause and effect analysis. The objective of a cause and effect analysis is to identify the ways by which a specific event, called a *top* or *undesired event,* can be caused, and to determine the effects of that event. For example, the undesired event may be the failure of the operator to receive an alarm of a particular adverse environmental condition. Having postulated this undesired event, the analyst would use a fault tree to represent the design and would reason inductively to identify the intermediate events leading to the undesired event, and by the same type of reasoning, subsequently arrive at the *primary* or *initiating event* for the undesired event. In a similar fashion, the analyst may begin with the postulation of a primary event, such as a component failure. Using an event tree, he may deduce the intermediate resultants of the primary event, leading to the deduction of the resultant top, undesired event. In either case, using inductive or deductive reasoning and working through the event gates, the analyst should identify any single failure event or any series of failure events that can cause the undesired event.

Now using an effects analysis, if it is found that the effects of the undesired event are unacceptable and if the probability of the undesired event is also unacceptably high, either the effects of the event have to be ameliorated or the probability of the occurrence of the event has to be reduced. To reduce the probability of the occurrence of the event, the design has to be adjusted (a) to eliminate any single failure that can cause the event or (b) to reduce the cumulative probability of the series of failures that can cause the event. Either redundancy has to be introduced into the design whereby the functions of a component could be performed by a second component in the event of the failure of the first (for case (a) above), or the reliabilities of the components have to be increased (for case (b) above).

The approaches to cause and effect analysis are shown in Fig. 6. Sometimes this analysis is referred to as probabilistic risk analysis or fault tree or event tree analysis. The detailed techniques for performing this analysis are not within the scope of this book. The purpose is merely to give the environmental manager an awareness of the availability of this technique. The specifics of this technique are covered in good books on product quality. This technique is also being used in the defense, space, and nuclear power industries.

The effects of a component failure may vary significantly depending upon the characteristic of the component that fails and its failure mechanism. For this reason another analytical technique, namely failure modes and effects analysis is often used to supplement the cause and effects analysis described above. Failure modes and effects analysis also can be used independently of cause and effect analysis. The

DESIGN ENGINEERING PROCESS

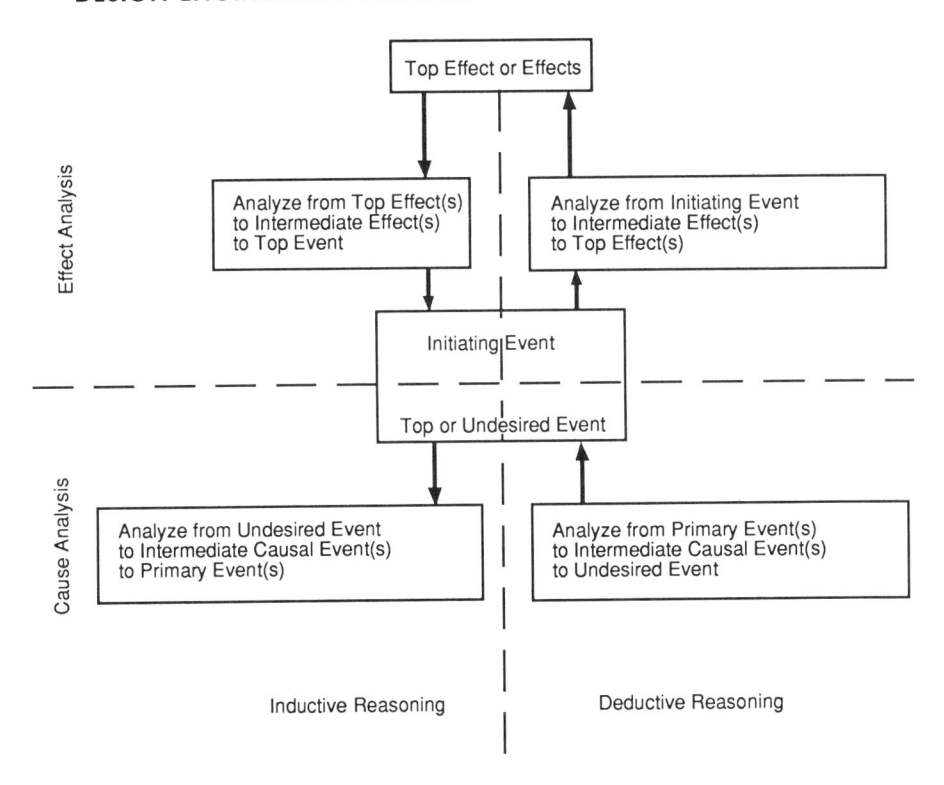

Figure 6 Approaches to cause and effect analysis.

objective of failure modes and effect analysis is to identify each possible mode of failure for each characteristic of the component in question and to determine the potential effect of the failure in each mode—with the further ultimate objective of causing the design to be altered to either minimize the failure mode's effect or to reduce the frequency of the failure mode's occurrence or both. Here again, the details of this analytical technique are beyond the scope of this book and the reader is referred to good books on product quality.

There are analytical techniques by which to assess the reliability and maintainability of items quantitatively, the ease with which items may be maintained, and the safety factors or margins of safety designed into the items (essentially physical or electrical load bearing capabilities compared to expected loadings). Each of these analytical techniques and more is intended to improve the reliability and maintainability of the items. There is even an analytical technique by which to assess the potential for failure which may be induced by the incompatibility of the equipment design with the limitations of the human being who must interface with the equipment in its operations—referred to as human factors analysis.

The management system should specify the conditions under which these analyses must be accomplished as part of the design process, with the analytical results becoming a part of the design review. Of course, the responsibilities for performing these analyses, the methods by which they should be performed, the timing of the completion of the analyses, and the contents of the analytical reports should also be specified procedurally.

Finally, the management system procedure should specify the administrative mechanics of the design reviews—for example, the timing of the design review relative to the design release; the information that is to be available to each design reviewer and when it is to be available; the form to be taken in the review process, the reviews done separately or jointly in a synergistic review meeting; the means of identifying action items resulting from the design review, usually through design review minutes; and the means of communicating to the appropriate levels of management, for their overview, the engineering decisions relative to major trade-offs (for example, the trade-off between equipment capability and equipment acquisition cost).

The environmental department, through its participation in the requirements and detailed design review and through its management overview of the trade-off decisions, can have a significant voice in helping to assure the adequacy of the design of the environmental hardware controls in addition to the environmental management system controls, both of which are necessary in the total environmental control program as discussed in Chapter 1 and depicted in Fig. 1.

6

Permit, License, and Approval Applications

INTRODUCTION

In general, regulatory permits provide authority to enter into construction of a facility whereas regulatory licenses provide authority to enter into operations of the facility. Sometimes, other approvals are required which are neither permits nor licenses per se, but which nevertheless constitute prerequisites to construction or operations.

Not only may these permits, licenses, and approvals impose constraints upon the start of construction or operations, but they may also impose constraints on how the construction or operations will be performed. As noted earlier, these constraints, once accepted by the company in consideration of the granting of the permit, license, or approval, have the same force of law as legislation, rules and regulations, and administrative and judicial decisions.

For any large construction or operating project there will be a variety of applicable permits, licenses, and approvals. They may vary, first of all, as to their source, emanating from the federal, state, county, city, township, and even smaller local levels. Within a single governmental level they may vary as to the regulatory agency which has jurisdiction. For example, at the federal level, it may be the Department of Energy, the Environmental Protection Agency, the Federal Energy Management Administration, the Federal Energy Regulatory Commission, the Nuclear Regulatory Commission or others. In addition to variations attributable to governmental levels and regulatory agency jurisdictions, the permits, licenses, and approvals may vary considerably with regard to the type of technical and administrative requirements imposed. Finally, the variations may also apply both to the amount of lead time required for the regulatory agency to process the permit, license, or approval application and to the frequency with which the application must be renewed. Because of the large quantities and large varieties of required permits, licenses, and approvals, it is necessary to develop and implement a management system to help to assure that they are acquired and renewed on a timely basis and that their constraints are appropriate.

ACCOUNTING AND DEFINITION

The first step in the management system should be to categorize the areas from which the permits, licenses, and approvals derive and to assign for each area an appropriate organization which is to be responsible for identifying and defining each permit, license, and approval emanating from that area. For example, the environmental department may be responsible for identifying and defining the permit and license applications emanating from the federal and state levels. Each facility may be responsible for identifying and defining those emanating from the local governmental entities within which the facility resides.

In addition to simply labeling the permit, license, or approval required, the responsible organization also should define the conditions for which the permit, license, or approval constitutes a prerequisite to construction or operations activities. At this stage, the definition of the conditions need not be to the ultimate level of specificity but they should be of sufficient specificity to enable one either to know that the permit/license/approval is required or that additional research is required in order to make the decision as to the need for these documents. The definition of the conditions should not be such as to possibly mislead one to think that the permit/license/approval is not required when, in fact, it may be. Generalities that allow error on the nonconservative side are dangerous and unacceptable from the management viewpoint.

APPLICATION ORIGINATION

Having identified the required applications and the conditions under which they are required, the next step is to identify the organization responsible for initiating each application. It may not be the same organization as is responsible for complying with the conditions of the application; for example, the organization that initiates the preparation of the application may not be the one that is to refrain from the construction or operations activities pending approval of the prerequisite application.

Undoubtedly, one of the most difficult aspects of the process is getting the construction organization, especially an outside contractor or itinerant constructor, to recognize these constraints. The constraints must be delineated clearly and the construction personnel must undergo specific training in the constraints.

The management system must provide a mechanism by which to update the list of required permits, licenses, and approvals. Obviously, this must be tied to the review of new environmental legislation and rules and regulations and to the regulatory responses to the permit, license, and approval applications—as well as to results of any adjudicated issues. New requirements must be factored into the system. The main points are to assure that:

- Means exist by which to consistently review new legislation and rules and regulations.
- Organizations have been assigned to review these new laws.
- It is understood that a major objective of the review of these documents is to identify any new or revised permit, license, and approval requirements.
- Once identified, further means exist by which to assign responsibility for originating the required applications, for complying with constraints until the prerequisite application ap-

provals have been granted and, thereafter, for complying with the conditions of the approved applications.

A means should also exist by which to control changes to this accountability list, a sample of which is provided in Appendix 6.1

Now, based on the information that has been made available from these reviews, the management system must also provide for the preparation of schedules by which to help to assure that the applications are originated and processed on a timely basis considering the lead time involved for the approvals by company management and by the regulators. The schedules should be living schedules, up-to-date at any point in time. Someone should be assigned to manage the applications process, to identify any applications that are not progressing according to schedule and to facilitate corrective action. The schedule should be sufficiently specific to provide the timing within which to process the application through each major organizational milepost and there should be a means by which to identify the status of each application relative to its schedule. If the company processes relatively few applications, the scheduling and statusing technique may be relatively simple. If, on the other hand, the processing of applications and their renewals is a frequent activity, more sophisticated scheduling and statusing techniques and tools should be made available. One should recognize that delays in processing applications may result in delays in starting construction or operations at significant cost to the company. Certainly the management system should describe the flow process for the applications from one organization to the next.

Having identified the organization responsible for the origination of the application, the management system should further describe the flow process for the application from one organization to the next; the elements of information to be added to the application by each organization; the timing for each step in the flow process; and the required approvals, with a definition of the attestation being made by each approval signature. The management system should also identify each application for which renewal is required and a scheduling technique should be provided by which to trigger the renewal process. The same kinds of controls should apply to other applications and to reports which are required to be submitted periodically.

Appendix 6.1 Master List of Nonradiological Environmental Permits, Licenses, and Approvals

	Agency	Type of permit/license/ approval (P/L/A)	Phase of project at which P/L/A is required	Approximate lead time[a]	Department responsible for acquiring P/L/A
1.0 AIR					
1.1 Federal	FCC	Operating permit for microwave transmitting equipment	Prior to microwave operation	4 months	CSD
1.2 State	MAPCC	Site approval. To provide reasonable assurance of compliance with state rules. Act 348 of PA 1965, as amended	Prior to any onsite earth change	10–18 months total (4–6 months for agency review)	ED
	MAPCC	Permit to install: Permit for boiler, auxiliary boiler, precipitator, stack fuel oil facilities, ash handling, pollution control equipment, etc. Act 348 of PA 1965, as amended	Prior to construction of fuel burning or pollution control equipment	6–42 months total (4–12 months for agency review)	ED

Appendix 6.1 *(continued)*

Agency	Type of permit/license/ approval (P/L/A)	Phase of project at which P/L/A is required	Approximate lead time[a]	Department responsible for acquiring P/L/A
MAPCC	PSD review process. Demonstration that the new air emission source will not interfere with the maintenance of the ambient air quality std. Clean Air Act, PL 95-95, 40 CFR 51	Prior to construction of any kind for a source of air emissions	18–42 months total (4–12 months for agency review)	ED
	Emissions offset process. Demonstration that progress toward attainment of the ambient air quality std. will be made prior to operation of the new air emission source. Clean Air Act, PL 95-95, 40 CFR 51	Prior to construction of any kind for a source of air emissions	24–42 months total (8–12 months for agency approval)	ED

	MAPCC	Stack height approval to assure stack height meets "Good Engineering Practices." Clean Air Act, PL 95-95, 40 CFR 51	Prior to construction of any kind for a source of air emissions	6–12 months	ED
		Approval of emissions and continuous monitoring/testing procedures	Prior to use	3–6 months	ED
	MAPCC	Permit to operate: boiler, auxiliary boiler, precipitator, fuel oil facilities, ash handling, pollution control equipment. etc. Act 348 of PA 1965, as amended	Apply not later than one month after unit begins operation	6–12 months	ED
2.0 WATER					
2.1 Federal	COE	Dredge and fill permit. Waters of the U.S. Clean Water Act of 1977, PL 95-217, Section 316 (b)	Prior to construction	4 months to 3 years	L&RWD

Appendix 6.1 (continued)

	Agency	Type of permit/license/ approval (P/L/A)	Phase of project at which P/L/A is required	Approximate lead time[a]	Department responsible for acquiring P/L/A
	COE	Work in, over or under navigable waters. Section 10, Rivers and Harbors Act of 1899	Prior to construction	4 months (minimum)	L&RWD
	USCG	Navigational aids. Buoys, lights, markers, etc.	Prior to construction	2–3 months	L&RWD
	USCG	Construction over navigable waters. As defined in Section 10, COE Title V, Section 502(b), General Bridge Act of 1946	Prior to construction	4 months (minimum)	L&RWD
2.2 State	DNR/EPA	Cooling water intake design approval. Clean Water Act, PL 95-217, Section 316 (b)	Prior to construction	18–30 months (to gather information) 12 months (for agency review)	ED

DNR/EPA	Thermal discharge demonstration (allows request for less stringent limitations). Clean Water Act, PL 95-217, Section 316(a)	Prior to construction	18–30 months (to gather information) 12 months (for agency review)	ED
DNR	Approval of structure in high risk erosion and flood risk area. Act 245 of PA 1970	Prior to construction of a permanent structure	4–8 months	ED
DNR	Certification of meeting water quality standards. Clean Water Act, PL 95-217, Section 401	Prior to construction (required before certain federal permits can be issued)	6–12 months	ED
DNR	Mineral well permit (test wells). Act 315 of PA 1969 (not needed if the driller already possesses a valid permit)	Prior to any drilling or well boring for water quality monitoring test wells	1–3 months (for agency review)	ED

Appendix 6.1 (continued)

Agency	Type of permit/license/ approval (P/L/A)	Phase of project at which P/L/A is required	Approximate lead time[a]	Department responsible for acquiring P/L/A
DNR	National Pollutant Discharge Elimination System Permit (establishes effluent std. for point source discharges). Clean Water Act and Act 245 of PA 1929, as amended	Prior to any discharge	8–24 months (for agency review)	ED
DNR	NPDES. Preoperational discharge. Clean Water Act of 1977, PL 95-217	Prior to any discharge during construction	4–12 months (for agency review)	ED
Health Dept.	Construction dewatering well installation. Act 294 of PA 1972	Prior to any well drilling	4 months	ED/BSD
DNR	Construction dewatering discharge approval	Prior to any discharge/pumping	4–12 months (for agency review)	ED/BSD

DNR	Natural Rivers Act. Activates in a designated "natural river area." Act 231 of PA 1970	Prior to construction of any kind	4 months	L&RWD
DNR	Floodplain Control Act. Activities in a floodplain area. Act 245 of PA 1929, as amended by Act 167 or PA 1968	Prior to construction	4 months	L&RWD
DNR	Work in or under water. Inland Lakes and Streams Act, Act 346 of PA 1972	Prior to construction	3–4 months	L&RWD
DNR	Wetlands Protection Act, Act 203 of PA 1979	Prior to construction	4 months	L&RWD
DNR	Great Lakes Submerged Lands Act. Use of lake bottomland. Act 247 of PA 1955, as amended	Prior to construction	4 months (minimum) (could exceed 2 years)	L&RWD
DNR	Permit to occupy bottomlands of Great Lakes. Act 10 of PA 1953	Prior to construction	2–4 months	L&RWD

Appendix 6.1 (*continued*)

Agency	Type of permit/license/ approval (P/L/A)	Phase of project at which P/L/A is required	Approximate lead time[a]	Department responsible for acquiring P/L/A
DNR	Dam construction permit (dam integrity approval) (impound water over 5 ft or area over 5 acres). Act 184 of PA 1963, as amended	Prior to construction	2–4 months	L&RWD
DNR	Water usage for pipeline hydrotest, gas transmission lines	Prior to construction	2–4 months	ED
DNR	Pollution incident prevention plan approval. Rule 323.1162, WRC General Rules	Prior to application for NPDES Permit	3–6 months (for agency review)	ED
2.3 Local County	Natural Rivers Act. Activities in a "designated natural river area." Act 231 of PA 1970	Prior to construction	1–3 months	L&RWD

3.0 LAND

3.1 Federal	FAA	Review of structure heights, stack, transmission lines, cooling towers, etc. 14 CFR 77	Prior to construction	3 months (minimum)	L&RWD/BSD
	FAA	Obstruction markings of tall structures. AC-70/7460-1F	Prior to construction	3 months (minimum)	L&RWD/BSD
3.2 State	Dept. of State. Michigan History Division	Coordination of gas transmission pipeline routing with state archaeologist	Prior to any construction in area	1 month	GEC
	DNR	Coastal zone management. Consistency certification. Federal Coastal Zone Management Act, PL 94-370	Prior to any construction in area	Agency has 6 months to challenge a certification; if challenged, resolution could require 1–2 years	ED
	DNR/Interagency	Soil erosion plan approval. Act 347 of PA 1972	Prior to construction	1–3 months	L&RWD

Appendix 6.1 *(continued)*

Agency	Type of permit/license/ approval (P/L/A)	Phase of project at which P/L/A is required	Approximate lead time[a]	Department responsible for acquiring P/L/A
DNR (District Fire Supervisor)	Permit to burn, e.g., site preparation materials. Act 329 of PA 1969	Prior to burning	1 month	L&RWD
MDOT	Permit for road alterations, crossings, driveways, etc. Act 368 of PA 1925	Prior to construction	1 month	L&RWD/BSD
MPSC	Permit to construct electric transmission lines. MPSC Order 1868 as revised	Prior to construction	1 month	L&RWD/BSD
MPSC	Permit to construct gas transmission lines. Commission Order	Prior to construction	1–4 months (minimum)	GED

3.2.1 Structures	MDOT	Railroad overhead wire crossing permit. Executive Order 1975-10 and MPSC Order 1868.	Prior to construction	1–3 months (minimum)	L&RWD
	MDOT	Highway crossing permits	Prior to construction activity within state highway right of way limits	1 month	L&RWD/BSD
	MDL	Permit to install: boiler; auxiliary boiler. Act 290 of PA 1965	Prior to construction	TBD	PMO/BSD
	MDL	Certificates for elevator use. 1965 Department Rule 408	Prior to elevator operation	TBD	PMO/BSD
	MDL	Building plans approval, barrier free variance. Section 6 of Act 230 of PA 1972	Prior to construction	3–6 months	L&RWD/BSD
	State Police	Construction approval, fuel tanks layout, above ground storage of flammable liquids. Rule 28.461	Prior to construction	3 months	PMO/BSD

Appendix 6.1 (*continued*)

Agency	Type of permit/license/ approval (P/L/A)	Phase of project at which P/L/A is required	Approximate lead time[a]	Department responsible for acquiring P/L/A
MAC	Obstruction marking of tall structure heights review. AC-70/7460-IF	Prior to construction	3 months	L&RWD/BSD
MAC	Height plans review, structure heights review. Act 259 of PA 1929	Prior to construction	1–3 months	L&RWD/BSD
3.3 Local				
County Road Commission	Natural beauty roads. Use of right of way of "designated natural beauty road." Act 150 of PA 1970	Prior to construction	2 months	L&RWD
County	Soil erosion plan approval. Grading or fill to protect sedimentation in waters of the state. Act 347 of PA 1972	Prior to construction	1–3 months	L&RWD/BSD

County	Permits for building, electrical, plumbing, mechanical. County ordinance	Prior to construction	1–3 months	L&RWD/BSD
County Road Commission	Permit for road alterations, crossings, driveways, etc. County ordinance	Prior to construction	1–3 months	L&RWD/BSD
County Drain Commission	Approval of drain crossings. Act 40 of PA 1956	Prior to construction	1–3 months	L&RWD/BSD
County Drain Commission or City, etc.	Surface water review. State Drain Code, Sections 20 and 22	TBD	TBD	PMO/BSD
County	Zoning requirements. Act 184 of PA 1943, as amended	TBD	TBD	L&RWD/BSD
County	Permit to set poles or structures in county road right of way	Prior to construction	1–2 months	L&RWD
County	Temporary guard pole setting and construction wire crossing of county roads	Prior to construction	1–2 months	ETEC (Construction Division)

Appendix 6.1 (continued)

Agency	Type of permit/license/ approval (P/L/A)	Phase of project at which P/L/A is required	Approximate lead time[a]	Department responsible for acquiring P/L/A
Township	Zoning requirements. Ordinance	Prior to construction	1–2 months	L&RWD/BSD
Regional Planning and Zoning Commission	Land use authorization	Prior to construction	1–2 months (minimum)	L&RWD/BSD
3.3.1 Structures Airport Zoning Agency	Structure height approval. Airport zoning permit	Prior to construction	2 months	L&RWD/BSD
Railroad Company	Railroad side track construction or modifications	Prior to construction	3–6 months	L&RWD/BSD
Railroad Company	Railroad crossing (overhead or underground) permit for gas and electric transmission lines	Prior to construction	3–6 months	L&RWD
Township, City or Incorporated Village	Building permits	Prior to construction	3–6 months	L&RWD/BSD

78

4.0 SOLID WASTE/SEWAGE DISPOSAL

4.1 State	Region Solid Waste Management Planning Agency	Solid waste management plan consistency review. Act 641 of PA 1979	Prior to construction	4–6 months	ED
	DNR	Advisory analysis, solid waste disposal facility site review by state. Act 641 of PA 1979	Prior to submittal of construction permit application	2 months	ED
	DNR	Permit to construct solid waste disposal facility. Satisfy solid waste requirements during construction of project. Act 641 of PA 1979	Prior to operation of facility	4–8 months	ED
	DNR	Permit to operate solid waste disposal facility. Act 641 of PA 1979	Prior to onsite construction of facility	6–12 months	ED

Appendix 6.1 (*continued*)

Agency	Type of permit/license/ approval (P/L/A)	Phase of project at which P/L/A is required	Approximate lead time[a]	Department responsible for acquiring P/L/A
DNR	Orders for water use and sewage treatment during construction and operation of facility. Act 245 of PA 1929, as amended	Prior to water use	2–4 months	ED
4.2 Local County Health Department/ City/Township	Approval of sewage system. County ordinance	Prior to construction	2–6 months	PMO/BSD
5.0 GENERAL ENVIRONMENTAL				
5.1 Federal Lead Review Agency	Environmental report submittal leading to approval of EIS. National Environmental Policy Act, PL 91-190 (non-nuclear facilities)	Prior to construction	18–24 months (for agency review)	ED

	FERC	Exemption request for minor use of oil/natural gas. Industrial Fuel Use Act, PL 95-620	Prior to burning oil or natural gas in new facilities capable of consuming 50 MMBtu/hr or more, exemption is required. Also required in some existing facilities	TBD	Fuel Supply
5.2 State	Regional Planning and Development Commission	Water use plans approval. Section 208 of PL 95-217	Prior to issuance of NPDES (operation)	2–3 months	ED
	MERB	Environmental report submittal leading to EIS approval. Executive Order 1974-4	Prior to construction	9–24 months (for agency review)	ED
	MERB	Gas transmission pipeline approval	Prior to construction	2–4 months	GEC

[a]Indicated lead times refer to the period after the permit or approval application has been submitted to the appropriate approving body. For major projects, permits that could be secured in 2–4 months for simple elements of the project cannot be issued until the project, as a whole, is approved; thus, the longest lead time permit or approval can be controlling.

7

Property Tax Exemptions for Environmental Equipment

INTRODUCTION

This chapter addresses the factors to be considered in establishing a management system by which to acquire property tax exemptions for environmental capital equipment.

DEFINING THE EQUIPMENT

The first step in the procedure is to define the equipment that may potentially qualify for exemption from state and local property taxes. An equipment potentially may qualify to be classified as "environmental equipment" because a large or significant portion of its cost is attributable to the performance of an environmental function. Also, the equipment may potentially qualify to be classified as environmental equipment because the environmental function being served is technically very significant—regardless of the portion of the equipment cost attributable to serving that function.

Simultaneously, there should be a definition of the attributes of an equipment which qualify it as "capital equipment" as contrasted to an item which must be "expensed" on an annual basis. Usually, the accounting department is responsible for providing this definition.

Also, simultaneously, there should be a definition of the legal constraints upon an environmental capital equipment which would disqualify it from consideration for exemption from state and local property taxes. The definition of these constraints should be provided by the legal department in terms that can be understood by the engineering, operations, and environmental departments.

Usually, either singly or jointly, the engineering, operations, and environmental departments should be required by the management system to identify the equipment that potentially qualifies as environmental. These organizations then should compare this equipment to the capital equipment definition and legal constraints previously provided. Usually, this is done at the outset of the project as the design progresses. However, a procedure such as this can be developed and implemented even after the project has become operational.

For each piece of equipment that passes through these decision gates, the organizations should document the functional and cost factors which support its classification as environmental capital equipment.

COST ACCOUNTING

Obviously, if there is to be an exemption from property taxes, the cost basis of the equipment must be known. Therefore, the management system should specify or make reference to a cost accounting procedure by which to account for the cost of the equipment. This includes the cost of its design and design support activities; the cost of the procurement of its materials and parts and the procurement support activities; and the cost of fabrication, assembly, installation, inspection, and test activities and their support activities. If this procedure is being implemented in the absence of a cost accounting system, special procedural provisions need to be made by which to acquire and document this cost data. The same applies if this procedure is being implemented after the fact of design and construction—after the equipment has become operational.

If the engineering and construction are to be done by an outside firm, the procedure should call for the review of the outside firm's cost accounting techniques. The accounting department should participate in the review—to assure that the techniques will facilitate the acquisition of supportable data to justify the tax exempt status of the equipment under the laws of the state or local entity. Contractual arrangements to enable this review as well as to enable appropriate change to the outside firm's cost accounting techniques, as necessary, should be provided up front at no additional cost.

APPLICATION FOR EXEMPTION

Having identified equipment that potentially qualifies for property tax exemption, having culled the list to eliminate "expensed" items and other equipment which does not meet the legal definition, having provided a technical justification for each remaining equipment for which tax exemption status is desired, and having established and justified the cost basis for each such equipment—the next step in the process is to prepare the application for exemption. The same procedural controls apply to the property tax exemption applications as applied to the construction permit and operating license applications described in Chapter 6. Suffice it to say that the procedure should specify:

- The organization responsible for providing or completing each element of information on the application form
- The processing sequence of the form
- The timing of the application processing sequence
- A means by which to assure that the application is being processed on time and by which to obtain corrective action if it is not
- The minimum content of each element of information on the form
- The approvals required on the form and the meaning of each approval signature (i.e., a definition of the attestation which is made by means of each approval signature)
- The identification of the organization which is responsible for submitting the application and following up to assure its timely processing by the state or local regulatory agency

Scheduling and adherence to schedule in the processing of tax exemption applications is very important because many state and local governments have cut-off times within which the applications must be received by the government in order for the government to process and approve the applications for the forthcoming tax year. If the cut-off date is missed and if special work-around arrangements cannot be negotiated, the company stands to lose the tax exemption benefit for the whole year.

The procedure should identify the organization responsible for receiving and retaining property tax exemption certificates and for communicating to the company's accounting, engineering, legal, and environmental departments the outcome of the application process.

Finally, recognizing that during the engineering and construction phases of a project and even during the operations phase, by means of major modifications, either additional costs may be sunk into equipments which qualify for property tax exemption or new equipments may be introduced that qualify. The procedure should address the responsibilities and methods by which to obtain additional certificates to reflect the added costs.

8
Project Management of Environmental Studies

INTRODUCTION

Many environmental professionals are educated as scientists or engineers and have performed in scientific or engineering capacities and not in project management capacities. Recognizing that each environmental study can be viewed as a project, the purpose of this chapter is to impart some fundamentals regarding the management of projects and to describe the most salient points of a management system by which these fundamentals can be applied consistently.

The application of project management fundamentals to environmental studies is important because it is often the practice to obtain these studies from outside scientific and engineering consulting firms. The company's in-house environmental expert should manage the study project in a way such as to protect the company's technical, cost, and regulatory interface interests. Not only is a large amount spent on the environmental study itself, but an amount many times larger may have to be spent depending on the study results. Even worse, the company can find itself in technical dispute with the regulatory agency if the study is not managed properly. It is not sufficient for an environmental scientist or engineer to address himself exclusively to the scientific and engineering aspects of the project.

THE PLAN

The company's management system procedure should first address the questions relating to the selection of the project manager. Which organization is to be assigned responsibility for managing environmental study projects? Is it to be the environmental department, the engineering department, the operations department, or a project management organization? What roles do each of these organizations play either in managing the project or in providing project management support? For example, if an environmental study project is to be managed by the project management organization, what type of technical support are the other organizations required to provide? The types of technical support to which I am referring will become apparent as we work through the project management process.

The next step is to prepare a project plan. Again, the organization responsible for the preparation of the plan and the organizations responsible for various elements of input to the plan should be procedurally specified. The plan should cover:

- The study objective(s)
- The scope of the study
- The requirements with which the study must comply
- The kinds of support that must be provided by the facility so as to accommodate the activities necessitated by the study
- The impact of the activities to be performed during the study (such as data collection activities) on the construction or operaton of the facility
- The schedule for the study
- The study's cost

Let us address some of these elements of the plan in sequence.

Study Plan Objective

The study plan objective should provide a statement as to the decision(s) that are expected to be reached based on the results forthcom-

ENVIRONMENTAL STUDIES

ing from the study. Usually a study is performed to facilitate a decision as to whether or not special engineering, construction, or operations activities are required to meet environmental requirements and commitments. If the study objective is other than this, its need, to begin with, should be questioned.

Scope Statement

The scope statement should be approached from a number of different aspects. For example, one aspect of the scope may be the enumeration of the different types of data elements to be acquired. Another may be the geographical definition of the area from which the data is to be acquired. Still another may be in terms of the different types of analytical tools to be applied to the data. And yet another way of defining the scope may be by an enumeration of the different types of tasks to be accomplished. Often each of these methods is used in combination to better define the scope. On the one hand, the scope statement is not intended to be, nor can it be, a definition of the methods to be used. On the other hand, it should be sufficiently specific to preclude a misunderstanding to the extent that the organization performing the study may later request additional funds or time to fully achieve the study objective.

Statement of Requirements

The same applies to the statement of the requirements with which the study must comply. It should be sufficiently specific as to:

- The qualifications of the personnel who are to be employed in the study
- The type and quantity of data to be collected
- The methods for data collection
- The types of data processing and analyses to be performed
- The data processing and analytical methodology
- The substance and content of the analytical report
- Any support the study organization is to provide in defending the report before regulatory or judicial bodies
- The quality controls that are to be employed throughout each step of the process

Failure to provide requirements of sufficient breadth, depth, and specificity can also lead to the study organization's request for additional funding and time. Even worse, it can lead to a report from which no decisions can be made or from which the decisions will be unsupportable or faulty.

The plan should specify the kinds of support expected from the facility. For example, will the facility have to provide any data and if so, what type? Will the facility have to drill any wells? Will the facility have to provide boats with which to collect aquatic life specimens?

The impact of the data collection and facilities support upon operations of the plant should also be specified in the plan. For example, will the facility have to operate to a prescribed operational scheme which departs from normal operation, for some time period, in order to accommodate data collection? The estimated schedule and cost of the study project also should be provided in the plan considering all of these support and operational factors.

As a final element, and it has been left until last for special emphasis, the plan should provide a definition of the action criteria, that is, a definition of the conditions which, if found to prevail from the study results, will lead to action and the definition of that action. For example, if the study results demonstrate the presence of a specified substance in a specified amount or greater beyond a specified geographic boundary, the facility will be modified to provide a filtration system of a specified capability and the modification will be commenced by a specified date. As another example, if the study results demonstrate with 90 percent confidence that there is no difference in the number, sizes, and health of a fish species indigenous to a given body of water before and after facility operations, the facility will not be required to provide any further deterrent to the loss of the species.

As stated in Chapter 2 dealing with environmental policies, and it bears repeating here, the major purpose for incorporating action criteria into the plan is to facilitate an agreement with the regulatory agency or community as to the threshold beyond which the company is to take action and the nature and schedule for that action, or within which the company is justified in taking no further action beyond the study itself. It is far easier to agree on threshold levels and types of actions before the fact that it is after the study results have been

ENVIRONMENTAL STUDIES

obtained, when emotions may run high and when judgments may be swayed by the study producing outcomes which were unexpected or insupportive of one preconceived position or another. It does not make much sense for a company to engage in a study which will embroil the company in conflict with the regulator or the community. The study results should provide a means by which to resolve the conflicts but that can only be if the study results are applied to preestablished action criteria.

The management system procedure should also address itself to the required reviews and approvals of the plan. The levels of approvals may be based on:

- The cost of implementing the plan
- The potential cost of implementing corrections should the study results indicate the needed for corrections
- Whether or not external agencies or the public are involved with the study project
- Whether or not the study project has a regulatory or community sensitivity

REQUEST FOR PROPOSAL

In addition to detailing the scope of the study project; the requirements with which the study project must comply; and the schedule, including schedule mileposts; other requirements should be procedurally covered and incorporated into the request for proposal (and ultimately into the contract) if the study project is to be performed by an outside scientific or engineering consulting firm.

One such requirement should address the provisions by which the study data and study report are to be kept secure and under the control and ownership of the company, not the outside firm. The release of the data or report prematurely to the regulator or to the public could severely damage the company's position from a number of viewpoints. First, the data may be erroneous and the company would not have the opportunity to correct the error. Once erroneous data is issued, regardless of its fallacious nature, it is difficult to correct the false impressions left in the public's mind. The data may be correct but the conclusions drawn from the findings may be incorrect. Also, the data and conclusions may be technically correct, but the language may be inflammatory.

Finally, even if the findings and conclusions are correct, it is in the best interests of the company to issue such a report to the regulator and public in conjunction with the company's plan for corrective action, assuming that action is needed. This should be done in a timely manner because the regulator and the public have a right to the information on a timely basis and because the company's action plan tends to dampen the adverse impact of a negative conclusion. As a corollary, there should be a contractual provision to the effect that the company has the opportunity to review and accept the study report before its issuance. This provision is needed for all the same reasons noted above—to correct data errors and conclusions, to eliminate editorialisms and insensitive language, and to incorporate corrective actions so as to demonstrate responsiveness.

Along a somewhat different vein, there should be a contractual agreement as to who owns the data. If the scientific or engineering consultant owns the data, the possibility exists that the consultant will present technical papers relating to the study project and that the company may suffer adverse publicity from such widespread distribution of the information. The company is obligated to prevent and correct its problems, but it is not obligated to advertise them beyond the regulatory and community boundary.

Another contractual provision should be that the consultant provide periodic progress reports by which to assess the technical, schedule, and cost adequacy of the work. While the company's study project manager should not depend on periodic reports for too much of his information, certainly these reports, if required to contain the proper information, can be of help in identifying problems on a timely basis such as to afford an opportunity for recovery without permanent adverse impact. Of course, the company's study project manager should be using other techniques on a day-to-day basis by which to assess technical, schedule, and cost performance. Among these are frequent interface with members of the consultant's project team at different management levels; review of the consultant's raw data using technical experts within the company; and for large study projects, even a third party audit or overview. Bad news early is good news.

THE PROJECT MANAGER'S RESPONSIBILITY

The management system should assign to the project manager at least the following responsibilities:

- To coordinate the interfaces among the various organizations participating in the study
- To assess the technical, schedule, and cost performance of the organization performing the study
- To assess the need for redirecting the scope of the study or the methods by which the study is being performed
- To be aware of unanticipated findings of significance and to communicate these findings to the proper personnel
- To assess the need for immediate corrective actions and to assure that these corrective actions are committed to and implemented by the appropriate personnel

All of these things must be done on a timely basis. Within these major responsibilities there are a host of activities which a project manager must perform. For example, he must:

- Remove schedule road blocks
- Arbitrate issues
- Keep communication channels open
- Promote esprit de corps, productively, and conformance to procedures
- Maintain knowledge of the project details and simultaneously maintain the big picture

STUDY REPORT

The management system procedure should require that the draft study report be reviewed for acceptance by the project manager, the environmental department, and each affected department. As a minimum, the review should assure that:

- The report correctly describes the scope, requirements, and methods used in the study.
- The findings are technically accurate and supported by scientific data.
- The conclusions are reasonable, based on the findings

- The report does not include inflammatory or insensitive statements.
- The report includes company corrective actions.
- The company has no substantial objections to seeing any part of the report, taken out of context, reprinted in the public media. (The fact that a finding is unfavorable is not a substantial objection.)

Prior to its final issuance, the project manager and environmental department should be required to review the report with the regulatory agency to assure that the valid needs of the agency are equally served by the report.

9

Environmental Agency Inspections

INTRODUCTION

The purpose of this chapter is to describe the factors that should be considered in establishing a management system by which to handle inspections performed by environmental agencies. The success of any inspection from both the company and agency perspectives is largely dependent upon the way in which the company interfaces with the agency inspector or inspection team. The benefits of a good management system accrue to both the company and the agency.

FACTORS TO BE CONSIDERED

First, at each facility, an individual should be procedurally designated to be the main interface with the agency inspector or inspection team leader. Also, an alternate should be designated to fill this role in the absence of the primary designee. The designation of a facility inspection coordinator and his alternate is important because often agency inspections are performed without any prior notification to the company. Often the agency's intent to perform an inspection is first known when the inspector(s) arrives at the company's facility. In such cases it would be difficult to establish a facility inspection coordinator on a case-by-case basis. In addition, the coordinator has a number of important tasks which will be described shortly and it is usually best that one person become expert in the performance of these tasks rather than having different people perform the tasks with different levels of expertise. The tasks require a strong sense of interpersonal relations as well as a strong technical and managerial sense. Valuable sensitivity in dealing with agency concerns can be gained by having one person perform in this capacity, inspection after inspection.

In addition, for inspection visits for which the company has received prior notification, the procedure should specify that the information received via the notification be transmitted to the facility reception and security personnel and that, upon the arrival of the inspector or inspection team, the information previously obtained be verified. If there were no prior notification, appropriate information should be obtained. In either event, the information procedurally required to be available upon reception should be as follows: each inspector's name and job title, the agency he represents, the expected duration of his visit, the name of the company employee with whom the inspector usually has contact, and a very brief explanation of the purpose of the inspection. In addition the procedure should require the reception or security personnel to ask for the credentials of each inspector, and should specify the type of documentation which is to serve as credentials. For each agency from which environmental inspectors may come, the company should know the documentation which the agency issues to its representatives. The reception or security personnel should be authorized to admit only those persons whose credentials are in order.

The procedure should specify that upon the arrival of an inspector or inspection team, the very first step to be taken by the reception or security personnel is to notify the facility inspection coordinator. The procedure should specify his name and telephone number and his alternate's name and telephone number as well.

The facility inspection coordinator has a number of important tasks which should be enumerated in the procedure. He should arrange for suitable office facilities for the inspector(s). He should notify the affected line organization and the environmental department. He should greet the inspector(s) on behalf of the company and explain his role as a facilitator—someone who is: to help the inspector(s) get accurate information on a timely basis from the company personnel who have both the responsibility and the knowledge for the subject in question; to start to acquire the documentation for the inspection; to understand the issues resulting from the inspection so as to start the process of clarification and corrective action as soon as possible; and, generally, to be useful to the inspector(s). The facility inspection coordinator should acquire more specific information as to the purpose of the inspection and the areas, processes, or documents to be inspected. With that information he should further alert the facility personnel who are likely to be contacted by the inspector, giving the facility personnel some insight as to the nature of the inspection to help them to prepare. In the case of an agency inspection team, he should organize a company team of coordinators from the affected line and functional organizations, such as to provide a coordinator for each inspector. Each of these coordinators should perform the same role for and with his agency inspector counterpart as the overall facility inspection coordinator performs for and with the inspection team leader.

While the procedure should specify that any coordinator is to assure that the inspector is led to the proper company personnel (i.e., in each case, the person who is both responsible and technically knowledgeable), the procedure should also require that when an inspector contacts an employee, as a company representative, the employee should answer only for his area of responsibility and only if he is sure that his answer is correct. Otherwise, he should refer the inspector to the proper source of responsibility and knowledge. The wording of this procedural requirement should be such as to not intrude on any employee's rights as a "whistleblower." When the employee is the responsible person, but when he is unsure of the correct answer, he should be instructed to tell the inspector that he will acquire an answer within a specified time.

Getting back to the role of the coordinator, he should be procedurally required to assure that if the inspector takes a sample of any air, soil, water, vegetation, animal life, or process material, the sample is shared as equally as possible between the agency and the company.

This is called a *split sample*. The purpose of a split sample is to try to assure that the sample to be analyzed by the agency and the sample available for analysis by the company are the same. Should the agency's analytical results indicate a problem, the company would want to be assured that the problem is with the sample itself and not with the analytical technique used by the agency. Furthermore, the coordinator should be responsible for assuring that the company's technical expert is present when the sample is taken. The expert, in turn, should be procedurally required to assure that the sample is properly taken and properly labeled and that, if the sample has a special storage condition or a limited storage life within which the analysis should be accomplished, these too are included on the label and made known to the inspector.

The coordinator should assure that the company's portion of the split sample is provided to the proper company personnel. The coordinator should provide a recommendation to the affected line or functional organization as to whether or not the company's portion of the split sample should be analyzed immediately. Obviously, for a sample with a limited life, there is no choice other than to perform the analysis, lest the company accept being more closely committed to the agency's results. This may be fine if it is easy to take another sample and process it quickly, or if that which is sampled is other than of major significance. Otherwise, the sample should be analyzed. In any event, the authority to analyze the sample should be given to the affected line or functional organization and to the environmental department—either organization independently having the option to analyze because either organization can be affected adversely by the outcome of the inspection and because each may have a different perspective.

The procedure should require that each coordinator record the noncompliances and open times resulting from the inspection and that these be communicated periodically and frequently to the overall facility inspection coordinator. The overall coordinator's responsibility should be to assure that the information is known to each affected line or functional organization, the environmental department and, in the event of the potential for a serious legal consequence, the legal department as well. The objective is to enable the organization responsible

ENVIRONMENTAL AGENCY INSPECTIONS

for corrective action or for providing the requested information to act on a timely basis.

The procedure should address the question of corrective action commitments, specifying those who are authorized to make commitments, the limitations applying to these authorizations, the methods for making the commitments, and questions of a similar nature. Obviously, a commitment may be made only by one who has the line or functional authority for the operation in question, assuming that he also has the resources with which to fulfill the commitment. (Chapter 1 provides a discussion of line and functional authority. Chapter 12 provides a more detailed discussion on corrective action commitment.)

One of the most important functions of the facility inspection coordinator is to try to promote the closure of open items and the initiation of corrective action before the inspector leaves the facility. The inspector may have a question or he may request documentation. Hopefully, the question can be answered and the documentation can be provided before his departure. The answer should be documented and a list of documents supplied to the inspector should be kept. This may help to identify the direction of future inspections.

Similarly, the inspector may identify a noncompliance and the corresponding corrective action, at least the corrective action to resolve the immediate condition at hand, should be provided to the inspector prior to his departure. If this cannot be done, it should be provided to him at least prior to the time that he will discuss his findings with his superior and finalize his written inspection report. The more that the report tends to stand alone, devoid of any open questions and devoid of noncompliance for which there are not corresponding corrective actions, the more the report tends to serve the valid interests of all those who have a stake in the outcome of the inspection. Open questions and open noncompliances sometimes can lead to unwarranted suppositions and misconceptions. Also, the responses to questions and noncompliances provided to the inspector may help to promote a more favorable overall tone of the inspection report.

Of course, the more complete the statement of corrective action provided to the inspector, the better. There are four types of corrective action that should be procedurally specified:

- The action to correct the noncompliance immediately at hand
- The action to prevent the recurrence of an identical or similar type of noncompliance in the process in which the noncompliance at hand occurred
- The identification of other processes in which identical or similar noncompliances may occur and the action to prevent these noncompliances
- The action to ameliorate the effects of the condition at hand pending the implementation of the first three corrective actions.

As a simple example, assume that a water sample was not taken on a timely basis and that the root cause is that the procedure is mute on the sample timing and that the employee has not been otherwise trained. Of course, the water sample must be taken to address the noncompliance at hand, the procedure must be corrected, and the employee must be trained in the procedural correction. Additionally, however, other procedures should be reviewed to determine whether or not they, too, have procedural voids as to the timing of the sampling. If so, these procedures must be corrected and the employees trained regarding the corrections as well. Finally, someone may be assigned to determine if the missed water sample resulted in any technical problem and, if so, to act accordingly and to take future samples on a timely basis pending the correction of the procedures.

The management system should also specify the minimal elements of a corrective action commitment to an inspection agency. They are a statement of each of the four types of corrective action (to the extent that each is applicable) and for each type, the date by which the action will be taken and the date by which the action will be effective rendering the processes wholly in control. Sometimes the date by which the action is completed and the date by which the action yields effective compliance are different. If corrective action for the process in question or for any similarly affected process is not necessary, or if ameliorating corrective action is not necessary, the agency should be afforded a good understanding as to why this is so.

Chapter 12 addresses correctives from a broader perspective, not limited to corrective action directly in response to agency inspection findings.

Any corrective action commitments provided orally or in writing in response to the written inspection report should be approved by both the line and functional organizations having authority over the process in question and by the environmental department. The line organization's approval should signify that the corrective action is technically and economically sound, that the action will satisfy the requirement, that resources exist for the implementation of the action, that its schedule is attainable and that the signatory accepts accountability for these things. The approval of the functional organization and the environmental department should signify that the action will yield compliance with regulatory requirements and company standards within a reasonable time.

One of the most difficult things to handle is technical disagreement between the agency inspector and the company representative whereby the inspector believes there to be a noncompliance and the company representative does not, or whereby the inspector believes that the company's proposed corrective action is inadequate. Aside from trying to resolve such issues before the inspector leaves the facility or, at least, before the inspector finalizes his report, there are no hard and fast procedural prescriptions but here are a few considerations. The question arises as to whether or not the company should escalate the issue to the agency management. Escalating an issue may be appropriate:

- If there is the potential for a large civil penalty
- When the corrective action requires a large expenditure
- When the corrective action itself does not involve a large expenditure, but when it establishes a precedent that may later involve a large expenditure
- When the inspector's position is so ill-founded as to be professionally incompetent
- When the company wishes to demonstrate its reasonableness to the agency management by escalating the issue only to yield and by yielding to enter into a give and take posture with the agency

On the other hand, escalating an issue may not be appropriate when the issue is insignificant relative to the points just enumerated and when escalation would only tend to prejudice the inspector and the inspection agency with regard to future inspections. In such cases, in its formal responses to the inspection report, the company may wish to provide words to the effect that while the company is taking an action to provide a "betterment" of the condition originally found by the inspector, the company believes that the original condition was in compliance with regulatory requirements. This kind of a statement may demonstrate a willingness to cooperate with the inspector and the agency while, at the same time, tending to balance the record and to avoid any admission of guilt.

The procedure should specify:

- The organization that is to receive the inspection report
- The organizations to which copies of the report are to be distributed
- The organization that is responsible for making the distribution
- The organization that is responsible for preparing the response (or if not procedurally specified beforehand, the organization that is responsible for assigning the response preparation and the criteria to be used for making such assignments)
- The timing of the preparation of the response
- The organizations that are to review and approve the response and the attributes of the response to which their reviews and approvals apply
- The timing of the preparation of the response
- The signatory of the response transmittal
- The timing of the transmittal
- The organization responsible for tracking the foregoing process and for making the transmittal and its distribution to persons other than the addressee (such as to other interested agencies, legal representatives, community and public interest group representatives, and company personnel)

Of particular importance, the procedure should address itself to tracking the corrective action commitments. Either the overall regulatory affairs department or the environmental department should track the status of each corrective action. This subject is addressed in detail in Chapter 12.

10

Reporting to Environmental Regulatory Agencies

INTRODUCTION

The purpose of this chapter is to describe the factors that should be considered in establishing a management system procedure for reporting apparent noncompliances with regulatory requirements and for reporting other significant events to the regulatory agency.

Before addressing the factors, a few words of caution from a legal viewpoint. Sometimes, when time does not permit a sufficient analysis by which to determine with certainty whether or not a noncompliance with regulatory requirements is reportable to the agency, a report may be made to the agency in order to meet the reporting time constraint. In the absence of certainty that the report is not required, making the report is the prudent thing to do. In such cases, it avoids the potential for another noncompliance, failure to meet the reporting time constraint, and it contributes to a better relationship with the regulatory agency. However, it may be well for the company to adopt language in the reporting form itself to the effect that the noncompliance is only "apparent" at the time of this report and that the report of the apparent noncompliance does not constitute an admission of noncompliance against the interests of the company. If such a reservation is to be used in the reporting process, it need not and should not be done with obtrusiveness. It should not be a big deal. From this point forward in the chapter, wherever reference is made to "noncompliance," it should be understood to mean "apparent noncompliance" as well.

FACTORS RELATING TO NONCOMPLIANCE

First, the procedure should define each type of noncompliance event for which the line or functional organization should provide a notification. These noncompliance events could be spills or releases of specified substances beyond specified boundaries and beyond specified levels. Or they could be cases in which the facilities are operated without the specified environmental protection equipment. Or they could be cases in which a certain environmentally significant action took place without first notifying the regulatory agency, as specified.

These specifications would be given in the requirements-type documents described in Chapter 3 and, therefore, the definition of events for which notification is required could be provided by reference to the requirements-type documents.

The procedure should specify those to whom the line and functional organizations must report the noncompliance. As a minimum, the line organization management should be the first to be notified so as to enable proper and timely corrective action to get underway. Also, the environmental department should be notified in all cases. The procedure may stipulate that the legal and public affairs departments also be notified in all cases or in selected cases that meet preestablished, more severe criteria. The procedure should specify the individuals within the line and functional organizations who are responsible for providing the notifications and specify the individuals within the environmental, legal, and public affairs departments who are the recipients of the notification, along with the information as to how these recipients can be contacted.

Usually it makes sense to provide the notifications in a two-step process—the first being an oral notification followed by a written notification. Often an oral report is a regulatory requirement. The

REPORTING TO REGULATORY AGENCIES

procedure should specify the timing of the oral report—i.e., the elapsed time after the noncompliance has occurred or after it has been detected, within which the notification must be made to each recipient.

Likewise, the timing for the written notification and the form and minimal content of the written notification should be specified. For example, the written notification should identify:

- The requirement with which there was a noncompliance
- The actual condition that existed
- The duration for which the noncompliant condition existed
- The root cause of the noncompliance
- The corrective action to be taken (recognizing the four types of action discussed in the preceding chapter and in Chapter 12)
- The impact of the noncompliance on the facility and its employees
- The duration of that impact
- The impact of the noncompliance on the public
- The duration of that impact as well

The criteria for reporting noncompliances to the regulatory agency may be different than the criteria for reporting internally. Therefore, the procedure should specify:

- The conditions beyond which reports are required to be made to the regulatory agency
- The organization responsible for the preparation of the reports
- The format and content of the reports, much of which is the same as for the internal reports
- The organizations responsible for reviewing and approving the reports, and the features of the reports to which their reviews and approvals apply
- The review processing sequence and the review and approval time allotted for each review/approval
- The organization responsible for making the report transmittal and the elapsed time from the date of the noncompliance or from the date of its discovery within which the report must be transmitted
- The regulatory agency to which the report must be transmitted, along with the additional distribution

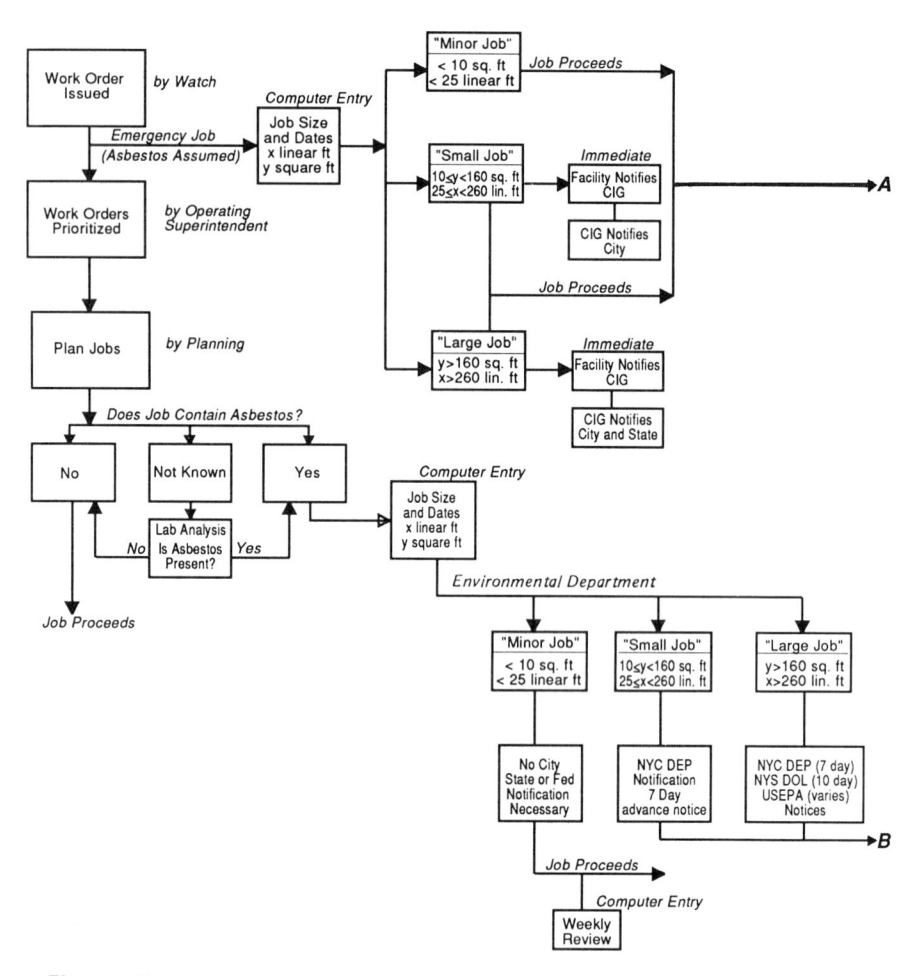

Figure 7 Asbestos notification tracking process flow.

Sometimes all of the information required to be contained in the report is not available within the time constraint for submitting the report. The procedure should address this situation to assure the timely acquisition, preparation, review, and approval and submittal of this information as well.

Usually, responsibility for making the final determination as to whether or not a report is required is given to the environmental department.

REPORTING TO REGULATORY AGENCIES

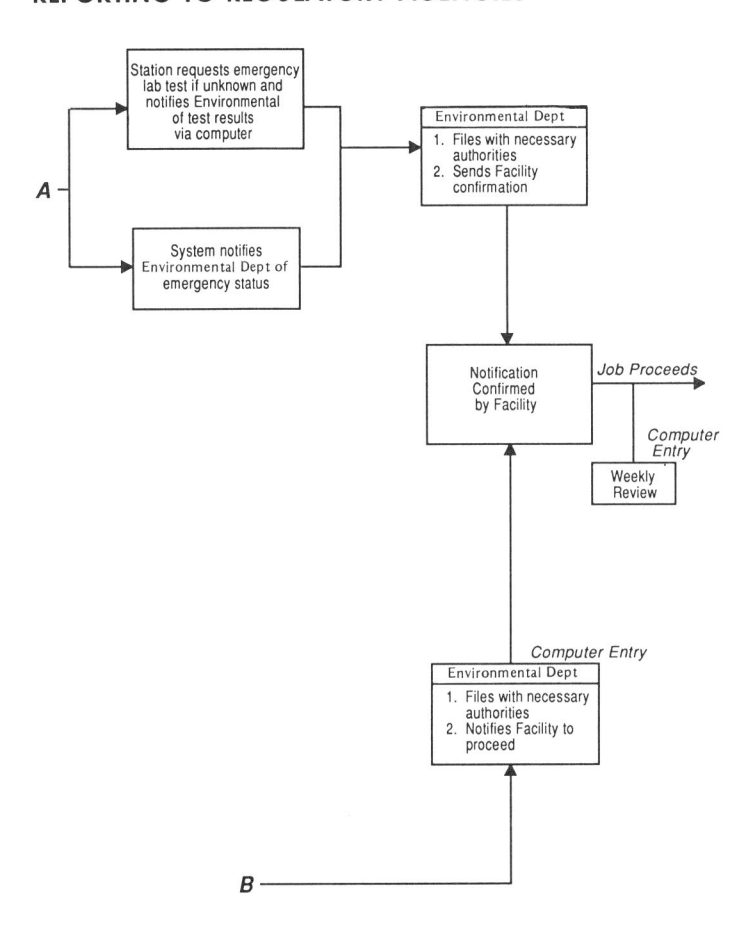

```
A ─┬─→ [Station requests emergency
   │    lab test if unknown and
   │    notifies Environmental
   │    of test results
   │    via computer]      ──→ [Environmental Dept
   │                            1. Files with necessary
   │                               authorities
   │                            2. Sends Facility
   │                               confirmation]
   │                                    │
   └─→ [System notifies                  │
        Environmental Dept of            │
        emergency status]                ↓
                                 [Notification          Job Proceeds
                                  Confirmed      ─────────────────→
                                  by Facility]
                                                     Computer
                                                      Entry
                                                  [Weekly
                                                   Review]

                                                 Computer Entry
                                 [Environmental Dept
                                  1. Files with necessary
                                     authorities
                                  2. Notifies Facility to
                                     proceed]
                                         ↑
B ───────────────────────────────────────┘
```

REPORTING SIGNIFICANT EVENTS OTHER THAN NONCOMPLIANCES

Sometimes it is required that significant environmental events be reported to an agency before the events occur. One such example is the removal of asbestos. The management system procedure should address the specific circumstances under which such notification is to be given, when it is to be given, who is to give it, and how it is to be given. For clarification, this may require a flow diagram, an example of which is provided in Fig. 7.

11

Environmental Audit

INTRODUCTION

The purpose of this chapter is to provide a discussion of the factors that should be considered in establishing management system procedures for both (a) environmental audits to be performed by the environmental department or some corporate entity and (b) environmental self audits to be performed by the line and functional organizations which are engaged in activities affecting compliance with environmental laws and commitments. This chapter addresses considerations such as:

- The scope of the environmental audit
- The selection of audit subjects
- Audit scheduling
- The planning of the audit
- The composition of the audit team
- The need for advanced notification of intent to audit
- The audit data collection process

- The treatment of audit findings
- The corrective action in response to audit findings (to a limited extent recognizing that Chapter 12 is devoted to the subject)

Before getting to these considerations, it must be understood that an environmental audit is only one element, albeit an important element, of a total environmental control program. Numerous times throughout this book the preventive nature of a good environmental control program has been stressed. In order for environmental noncompliance to be prevented there must be:

- Good policies
- Good requirements-type documents
- Good procedures by which to attain compliance with the requirements
- Good in-process measurements and assessments, in real or in reasonable time, by which to determine whether or not compliance is being attained
- Effective and timely corrective action
- Thorough training in all of the foregoing elements of the program
- Audit

Early in the environmental era, many companies believed that the environmental audit constituted a total environmental control program. Some companies still may have this belief.

SCOPE OF THE ENVIRONMENTAL AUDIT

In establishing the scope of the environmental audit basically there are two options. The first is to limit the scope of the audit to an assessment only of the degree of compliance with policies, requirements-type documents, and procedures. The second is to assess the adequacy of the policies, requirements-type documents, and procedures to begin with and, given their adequacies, to then assess the degree to which compliance with these documents is achieved.

The argument for limiting the audit scope only to the assessment of compliance is that the policies, requirements-type documents, and procedures have been developed and approved (ostensibly) by the key affected technical and managerial personnel. The value of further

THE ENVIRONMENTAL AUDIT

assessment of the adequacy of these documents is questionable. The auditors are not likely to possess any greater understanding than was possessed by the technical and managerial personnel. More likely, the auditors will raise issues that already have been identified and dispositioned by the proper authorities in the first place. An audit for adequacy will provide only flack for the opponents of the company's operations, the regulators who might oppose the company's earlier dispositions of these issues, and those within the company who, in the past, have tried to acquire different dispositions of the issues and now, with the backing of these so-called new audit findings, will try again. The bottom line is that much effort will be wasted revisiting old issues.

On the other hand, environmental laws change (mostly by additions these days) and community levels of sensitivity to environmental issues also change. The nature of the company's operations also may change such as to increase environmental risks in the absence of more stringent policies, requirements-type documents, and procedures. Technology and management systems advances possibly can provide low cost opportunities for improvement. Certainly under these conditions, an audit's reassessment of the adequacy of the policies, requirements-type documents, and procedures is warranted.

It appears that the audit objectives should be both to assess the degree of compliance with policies, requirement documents, and procedures and to assess their adequacy as well, with the following constraints. When a function, process or area is subject to frequent, periodic audit each audit need not address the adequacy of the environmental program. The large majority of the audits should address only the degree of implementation compliance with an occasional audit addressing initially both the degree of implementation compliance and the adequacy of the program. Furthermore, when the adequacy of the program is to be addressed, there should be a higher level overview by the management of the auditing organization—an overview of any findings related to programmatic inadequacy. The purpose of this overview should be to assure that each issue is fresh or if not, that at least there is sufficient cause, as described above, to warrant a revisitation of the issue before putting other organizations in the company through the perturbation of addressing these issues.

CHAPTER 11

SELECTING AND SCHEDULING AUDIT SUBJECTS

Basically there are two ways by which to select subjects for environmental audit. The first is to select an environmental management system, addressing the elements of that system regardless of the organization or geographic areas in which the system elements are accomplished. The idea here is to audit the system from its beginning step to its concluding step, recognizing that different steps are performed by different organizations at different locations. The emphasis in this case is on the system itself rather than on the organizations or geographic areas.

Conversely all of the environmental activities performed by a particular organization or performed at a particular geographic location may be audited regardless of the systems into which these activities fall. In this case, the emphasis is obviously on the organization's or the area's degree of compliance.

For a given period, environmental audits may be scheduled such as to cover all or a portion of the management systems or organizations/geographic areas. For example, it may be required that each management system procedure be audited within any given twenty-four month period. Similarly, it may be required that each facility be audited within any given twelve month period. The purpose of this kind of scheduling is to cover each procedure, organization, or geographic area routinely within a limited period so as to help to assure that any noncompliance does not continue beyond that limited period.

Environmental audits may also be scheduled as requested by management based on the need for some additional assurance. For example, if an activity is about to start which could significantly impact the environment, in the interest of prudence, management may request a special audit of the preparedness for that activity; or if it is suspected that a regulatory inspection may be made of a particular activity or area, management may request that an audit be performed of that activity or area in advance of the regulatory inspection; or if management suspects the inadequacy of an environmental procedure or suspects noncompliance with the procedure in any particular organization or geographic area, an audit may be requested to address the management concern.

THE ENVIRONMENTAL AUDIT

However, environmental audits should not be used to readdress a known problem. It makes little sense to waste audit effort on environmental inadequacies or noncompliances of known existence, degree, and effect. In addition, in the case of a request for such an audit, the requestor's intention may be to acquire an audit report as a "wedge" by which to make a point that he may have been unsuccessful in making in the past. In such a case, regardless of the auditor's sympathy with the point, the audit request should be denied, if possible, to help to preserve the audit's complete impartiality. Not only must the audit be impartial but it must be perceived to be impartial as well. On the other hand, it may be difficult to deny such a request when it comes from a member of senior management or a customer and, therefore, carries with it certain political implications which must be recognized.

In general, the greater the number of environmental audit requests from management, the greater the acceptance of the function.

PLANNING THE AUDIT

There are a few factors, the affects of which routinely should be considered before performing any audit. One such factor addresses the question of the scope of the audit, as was discussed at the beginning of this chapter. From a planning viewpoint, the audit scope bears upon the selection of the audit team members and the duration of the audit. Obviously, the broader the scope, the broader the capabilities and the longer the duration needed for the audit.

Another factor to be considered in the planning phase is whether or not to provide a notification of the intent to perform the audit significantly in advance of its performance. Generally speaking, it is best to provide such advanced notification so as to facilitate the interface of the auditors with the personnel in the organization being audited. It makes little sense to put the auditors in a position whereby they may have incomplete information due to the unavailability of key personnel in the organization being audited. This may delay the audit because of disagreement with the facts as will be discussed later in the chapter. This may also impart to the organization being audited a sense that the auditors are not trying to be helpful. However, if advanced notice could bias the audit's results significantly through unrepresentative behavior, the only reasonable plan is to perform the audit without giving advanced notification. The auditors have to project from a relatively small sample to a population and if the sample was biased, the projected conclusions about the population will be incorrect.

The composition of the audit team must also be planned. Sometimes special skills are necessary for the performance of an audit—skills that cannot be maintained economically on the audit staff on a full time basis. Under these conditions, the audit management system should allow for the formation of an audit team comprising personnel with the required skills. As a rule, the team members should be from organizations other than the one having primary responsibility for the function or the geographical area to be audited. Obviously, this is to promote objectivity and to prevent any possibility of real or imagined intimidation.

The audit plan should also provide a realistic completion date for the audit—one that will permit the reporting of current status. An audit conducted over a long period of time loses it significance in that questions will arise as to the current validity or applicability of results obtained during the early phases of the audit. The audit plan should strive for audit results which will not be stale.

The audit team should plan the process of data collection. Often, in the interest of economy and schedule, auditors must rely upon sample information from which to make findings. In these cases, statistically planned samples are a must, carefully avoiding samples sizes that are too small to yield decisive information. If the sampling errors are too large or if the statistical confidence is too low, it will be difficult to make a decision as to a finding. No decision is no audit.

During the planning stage, the policies, requirements-type documents, and procedures which provide the criteria against which to assess performance should be identified and from these documents audit checklists may be developed to help to assure the completeness of the audit.

PERFORMING THE AUDIT

Certainly one of the most important steps in the audit process is to provide the advanced notification when such is the plan. This notifica-

THE ENVIRONMENTAL AUDIT

tion should be provided reasonably well in advance of the start of the audit. As part of this notification, a pre-audit conference should be scheduled, the purpose of which is to introduce the auditors to the supervisory personnel in the area being audited, further discuss the audit objectives and scope, define the criteria upon which the audit will be based, discuss the data collection techniques, establish the interfaces between the auditors and the supervisors, and discuss the process by which the auditors will feedback information to the supervisors such that there can be early agreement on the facts and timely corrective action. In general, at this conference, emphasis is placed on techniques aimed at gaining "acceptance" of the audit process. There are many personal deterrents to the acceptance of an audit function and to the attainment of corrective action. There are at least six threats posed by the auditor*: the threat to dignity (we're sensitive about our ignorance); the threat to freedom of choice (we're already controlled to the hilt); the threat to group prestige (don't tell us how to run our job); the threat to security (an outsider made an improvement); the threat to personal ambition (who gets credit for the improvement); the threat to old habits (we've always done it this way).

There will be times when the auditor, regardless of his efforts to secure cooperation, will encounter excessive opposition as when, for example, an effort is made to withhold information or to delay the audit process. The auditor must be able to detect these situations and overcome them without using an authoritarian approach.

Audit findings should be clear and accurate. By clarity it is meant that the statement of the finding is given with such precision that it can be interpreted in only one way—the way in which the auditor intended. By accuracy it is meant that the statement of the finding is factual; that data support the statement; and that the statement is free of any editorial content, including conclusions as to the impact of the finding. Conclusions by their very nature are not factual. This is not intended to imply that the audit report should be devoid of editorial content, such as conclusions or the impact of findings. It is to say, however, that a clear line of demarcation is needed between the statement of the finding and the statement of any conclusion drawn from the finding.

---

*From a Hughes Aircraft Company pamphlet.

Findings may be of different types. For example, (a) policies, procedures, or requirements-type documents may be apparently inadequate or there may be apparent noncompliance with these documents, (b) conditions may exist which if allowed to continue may result in apparent noncompliance with these documents, (c) conditions may exist for which the compliancy status cannot be ascertained. Findings may be of different levels of severity. Conditions may be such as to require immediate work stoppage or notifications to regulatory agencies. Except in the case of imminent danger, the authority to stop work should be limited to line management. In all cases, notifications of apparent noncompliance should not be the responsibility of the auditors.

At the conclusion of the audit, there should be a post-audit conference during which the audit team leader should get agreement on the facts of each finding—agreement between the auditor and the manager of the function or area being audited. If agreement cannot be reached, possibly additional data should be collected jointly by the audit and line organization personnel. In addition to agreement on the facts of the findings, at this meeting the audit team leader should strive for agreement as to the significance of each finding and its root cause(s). Both of these elements of information are necessary for arriving at a corrective action commitment which should be the audit team leader's ultimate objective for the conference.

A *root cause* is a cause which when eliminated will result in the avoidance of a repetition of the noncompliance at hand, as well as the avoidance of a similar noncompliance in the process which was audited or in any other process. As a first step, the corrective action commitment should address the means by which to identify the root cause if it is not obvious. If it is obvious, the corrective action commitment can address the means by which to eliminate the root cause. In either event, at the postaudit conference the audit team leader should strive for a commitment which includes providing the following elements of information as a minimum:

- What is to be done?
- Who is to do it?
- When is it to be completed?

THE ENVIRONMENTAL AUDIT

- When will its completion render the management system or the process in compliance with the requirements?
- What ameliorating or compensating action is to be taken until then and who is to take it?

It is to everyone's benefit to have the corrective action information available for incorporation into the audit report. From the audit team's viewpoint, it demonstrates that the audit resulted or is resulting in concrete improvement and, as such, reinforces the technical and economic justifications for environmental audits to begin with. From the functional or line organization's viewpoint, the corrective action commitments incorporated into the audit report demonstrate the responsiveness of the audited organization by having provided a timely response and a reasonable limitation on the period of noncompliance.

AUDIT REPORTING

The environmental audit management system procedure should address the audit report format and contents. The report format should provide for statements as to:

- The scope of the audit
- The data collection method
- Each finding, which may include a description of the "as required" condition compared to the "as found" condition
- The past, present and future impacts of each finding
- The corrective action commitment for each finding, including the elements of the commitment discussed in the preceding section
- For each finding for which a commitment is not now provided, a statement identifying the actionee and requesting a corrective action commitment by a specified date (which should be procedurally established)

The question of the need for a management summary is a difficult one. The disadvantages of a summary may outweigh its advantages. In summarizing, significant points may be omitted which can alter the finding and give it a broader, more serious flavor. This can be misleading and objectionable to those responsible for the function or area being audited.

The audit procedure should address the required reviews of the audit report prior to its issuance, specifying higher levels of reviews for the elements of the report that address issues of policy and procedural adequacy.

The audit procedure should address the resolution of conflict or refer to another procedure which addresses that subject.

The procedure should cover the distribution of the audit reports. In general, the reports should be addressed to those who have responsibility primarily for the function or area being audited and to those who have the responsibility primarily for taking corrective action. Copies of the audit report should be sent to at least one level of management above the addressees and to the management of the environmental department. Copies of the report should also be sent to the public affairs and legal departments if the conditions found during the audit have impact beyond the company boundaries.

The procedure should address the method for closing out the audit findings.

The procedure should also address the treatment of audit reports and closure documents as official company records.

Finally, the procedure should address the issuance of status reports to management, on an organization-by-organization basis, which should provide information as to the corrective action requests for which commitment responses are overdue and the corrective action commitments for which the action is overdue—the purpose of this score card is to get management attention focused on the closure of the audit findings. An example of such a report is provided in Fig. 8.

Much more could be said about corrective action under the heading of environmental audit but to avoid duplication it will be covered in the next chapter, which is intended to address corrective action in depth.

ENVIRONMENTAL SELF-AUDIT

In more advanced organizations, in addition to environmental audits performed by a corporate environmental department, an environmental self-audit is performed by the organizations having various environ-

THE ENVIRONMENTAL AUDIT

Facility	Responsible department	Responsible person	Status code	Open item description (and audit reference no.)
A	1	a	2	
		b	3	
		b	7	
	2	a	1	
		a	4	
		a	8	
B	1	a	3	
		b	1	
		c	7	

Status Code Descriptions:
1. Original response date not met by responsible person.
2. Extended response date not met by responsible person.
3. Response unacceptable from a technical viewpoint.
4. Response unacceptable from a schedule viewpoint.
5. Response evaluation not completed by ED as scheduled.
6. Corrective action not completed by responsible person as committed.
7. Corrective action not completed by responsible person as recommitted.
8. Corrective action verification not completed by ED as scheduled.

Figure 8 Example of an environmental audit corrective action exception report.

mental responsibilities. This is especially beneficial for special process and facility operations which could have major adverse environmental impacts. All the questions applicable to corporate audits are equally applicable to self-audits and should be addressed procedurally.

12
Environmental Noncompliance Reporting and Corrective Action

INTRODUCTION

The objectives of an environmental noncompliance reporting and corrective action management system are:

- First, to assure that noncompliances are reported and that the noncompliant conditions are corrected or otherwise addressed, as necessary
- Second, to assure that the root causes of the noncompliances are identified and eliminated such as to prevent future noncompliances of the same or similar types
- Third, to assure that feedback of the frequencies of noncompliances and the status of corrective actions is given to management so as to facilitate the management role in noncompliance reduction and corrective action adequacy and timeliness
- Fourth, to assure that the dispositions of noncompliances are such as to demonstrate a disciplined professionalism, which not only contributes to a greater probability of making the right

disposition but also contributes to ameliorating the legal consequences when the right disposition is missed

ENVIRONMENTAL NONCOMPLIANCE

The management system should provide a definition of environmental noncompliance. There are at least four different conditions, as follows, which constitute environmental noncompliance.

Environmental noncompliance exists in any case for which a lower tier environmental program document is inconsistent or incompatible with a higher tier document—as, for example, when a procedure for reporting environmental noncompliance to a regulatory agency does not enable the report to be made within the legal time limitation, or when an operating procedure is inconsistent with a provision of the plant operating license. A policy, requirements-type document, or procedural inadequacy may be the most serious type of noncompliance because the noncompliance is systematically repeated with each implementation of the document.

A second type of environmental noncompliance occurs in any case for which there is a failure to follow the requirements of a policy, a requirements-type document, or a procedure—such as when a plant is operated in violation of an operating procedure, or when any environmental measuring device is not given preventive maintenance or is not calibrated in accordance with the required schedule, or when a notification of environmental noncompliance is not provided as procedurally required. If the company's environmental control program has matured beyond the levels of *insensitivity* and mere *awareness* to a level of *enlightment* or *certainty,* there should be a very low frequency of noncompliances of this type. At high levels of maturity, the company culture is such as to require absolute adherence to requirements in a management atmosphere which promotes participative contributions to the requirements to begin with, as well as training in the requirements and methods by which to change the requirements when change is necessary.

As a corollary, the third type of environmental noncompliance occurs in any case for which there is a failure of an equipment which is necessary for attaining or for assessing the attainment of environmental

NONCOMPLIANCE AND CORRECTIVE ACTION

compliance. Such a failure can occur without there necessarily having been any inadequacy (any noncompliance of the first type) or any lack of adherence to the requirements (any noncompliance of the second type). In other words, the equipment failure may not be attributable to a design inadequacy or to an operating procedure inadequacy (first type) or to manufacturing or installation error or to operating error (second type). The equipment simply may have worn out.

The fourth type of noncompliance occurs when the management control or design control is such that a determination cannot be made as to whether or not compliance is attained. The company must be knowledgeable of its state of compliance in real or in reasonable time.

Noncompliance should not be limited strictly to the regulatory arena. Nonregulatory environmental documents may be inadequate. For example, a management objective may be to choose a site for a new facility which is the most economical from the point of view of maximizing income and minimizing expenses, while, at the same time, attaining compliance with environmental law and company environmental commitments, or a management objective may be to obtain the maximum amount of tax exemptions that are legally allowable for environmental equipment. Obviously, if the procedures for environmental considerations in site selection or for environmental property tax exemption applications are inadequate or if they are not followed, certainly there exist noncompliances. The only difference is that these noncompliances may not be of interest to an environmental regulator.

IDENTIFYING, RECORDING, AND COMMUNICATING ENVIRONMENTAL NONCOMPLIANCE

The management system should specify those who are responsible primarily for identifying environmental noncompliance while, at the same time, giving others the opportunity to identify environmental noncompliances as well.

The form on which an environmental noncompliance is to be recorded should also be specified. In general, the form should provide for the following different types of information.

First, the form should be identified as an environmental report or an environmental open item report which is preferable to identifying it as an environmental noncompliance report. The document could become evidence in a legal proceeding. While the best intentions may be to provide a factual description of a condition which is truly noncompliant, inadvertently the report may be factually erroneous or it may provide nonfactual information mixed with factual information in a way that is disadvantageous to the company's interest. However, recognizing these possibilities, it is still in the company's best interest to have a report in which to identify environmental open items. How else could environmental problems be identified and corrected with certainty?

The report form should also provide for a unique identifier, either simply a number or an alphanumeric. The identifier may provide a codified identification of the facility to which the report applies, the year in which the report is being originated, and the sequential number of such reports originated for that facility in that year. The main purpose of the unique identifier, however, is merely to enable one to quickly distingusih one report from another.

The report should provide a section in which to identify the thing to which the report applies. For example: the location, the facility, the equipment, the part of the equipment, the document, the subsection of the document, or the activity that is nonconforming.

Another section of the report should identify the requirement. This may be a quotation from the law or a policy, but preferably from a requirements-type document or a procedure—along with the identification of the source from which the requirement is taken. If one has to quote the law or a policy because there is no requirements-type document or procedure to cover the subject, that, in itself, is a noncompliance. This section of the report should also state the existing or "as found" condition. The statement should be in factual terms and should be quantified, if possible. The form and the instructions for completing the form should be such as to emphasize that nonfactual information is not provided in this section. It is best that this section remain pure in this regard so as to be able to legally distinguish between the factual information in this section and nonfactual information which may appear in other sections of the report.

The next section of the report should address any secondary con-

NONCOMPLIANCE AND CORRECTIVE ACTION 125

ditions which result from the primary condition. For example, if a requirements-type document is inadequate in that it fails to specify a particular environmental measurement or if there is a failure to follow procedure and the required measurement is not made—either condition having been reported in the preceding section—an outflow of this could be that the environmental state of the process is unknown and, as such, conformance to environmental law or commitments cannot be assured. This is the kind of information that would be included in this section.

Notice that for both the primary and secondary conditions there is no provision at this stage of the form by which to document the significance or the seriousness of the impact of the conditions. This is something that may be well beyond the capability of the report originator to determine.

The next section of the form should identify whether or not the condition is required to be reported orally and, if so, to whom and when the oral report was actually made. This is in consideration of the possibility that, in turn, a report must be made to the regulatory agency.

At this junction, the form should provide an opportunity for the originator to provide an opinion as to the corrective actions that should be taken, who should take them, and when they should be taken. This is the only part of the form prepared by the originator that may be nonfactual. Each other part of the form should provide only factual information.

The final section of the form to be completed by the originator should identify the originator and provide his dated signature attesting to the accuracy of the information which he entered on the form. This section should also identify the reviewer of the form and provide his dated signature attesting to the form's completion in accordance with its completion requirements.

The management system procedure should address the timeliness of the origination and distribution of the form as well as identify the individuals to whom the form should be distributed. As a minimum, the distribution should be to the supervisor (at an appropriate level) of the organization which failed to take the action or which took the action which resulted in the noncompliance; to the next level management person in that organization; to a management person in the organization which may be impacted by a secondary noncompliance resulting from the primary noncompliance; to a management person in the organization which is responsible for taking corrective action or for taking the next step in the process of closing out the environmental open item report; and to the environmental department personnel who are responsible for overseeing the processes which were impacted by the primary and secondary noncompliances.

THE CONDITION AT HAND

As noted earlier in this chapter, the second objective of the management system should be to assure that the noncompliance at hand is either corrected or otherwise addressed. If the policy, requirements-type document, or procedure is inadequate, revise it to make it adequate. If the procedure is not being followed, follow it in the future. If the equipment failed, fix it. If the inadequacy, the failure to follow the procedure, or the equipment failure yielded a secondary noncompliance such as it not being known whether or not the environmental condition complies with its requirements, make the measurements or assessments and determine whether or not the condition complies. If it does not, fix it or otherwise address it.

Inadequacies in environmental policies, requirements-type documents, or procedures; failures to follow these documents; and failures of environmental equipment must always be fixed. Sometimes the resulting condition of the environment itself cannot be fixed. Sometimes the resulting condition of the environment, although noncompliant, does not warrant fixing because the cost of the fix relative to the cost of the risk is substantially out of balance. Usually this out-of-balance is attributable to the requirement having a substantial margin of safety while the noncompliant environmental condition has no significant contribution to a reduction in safety itself or to a reduction of the safety margin—i.e., the margin between the level in which safety is impacted and the level of the requirement.

The procedure should require a response to the environmental open item report. The response should be required within a specified time after the issuance of the report. The response should address both the noncompliant condition at hand and the impact of the condition. For the condition at hand, the response should state:

- The condition to be fixed
- When it is to be fixed
- Who is to fix it
- When the fix is to result in compliance
- What alternate method is to be used to maintain compliance until the fix is effective
- If no alternative technique is to be utilized and if operation is to be continued, what justification exists for continued operation

For the impact of the noncompliance, the response should state:

- What is to be done to measure or assess the condition of the environment resulting from the primary noncompliance
- Who is to do it
- When it is to be completed
- If the condition of the environment is already known to be noncompliant, what is to be done to make it compliant
- Who is to do it
- When it is to be completed
- If the noncompliant condition of the environment is not to be corrected, what justification exists for the absence of correction

The procedure should stress that responses are inadequate if they do not provide answers to each of these questions, as applicable. Furthermore, the system should stress that the justification for lack of action must be documented, technically adequate, and based on analysis—not simply seat-of-the-pants "engineering judgement." This is not to say that every condition requires a detailed analysis. A bounding analysis may serve in many cases.

In addition, the procedure should identify the individuals who have the authority to analyze the safety impact of a noncompliant environmental condition and who have the authority to make any decision to continue to operate with the noncompliant condition. The analysis must be quantified and the decision maker must be at a broad enough level of technical and managerial responsibility to understand the full safety, technical, legal, public affairs, regulatory, employee, and financial impact of any decision—especially a decision to take no action.

The environmental open item report form should be structured such as to provide space for the documentation, or the reference to the documentation, specifying (a) the corrective action to be taken for the primary noncompliance, (b) the corrective action to be taken for the secondary noncompliance (the impact of the primary noncompliance), or (c) the justification for the absence of such action. The proper signatures should be provided on the form or on the documents referenced in the form.

Next, the management system should address the closure of the primary and secondary noncompliances for the condition at hand. Should the environmental department be given the responsibility to verify that the action claimed to have been taken was, in fact, taken? Should the verification of the action be referenced in an environmental open item report? Should the method of verifying the implementation of the corrective action be stated in the report, along with the identification and dated signature of the verifier? If so, it would provide a logical and central close out of the report for the condition at hand.

ROOT CAUSE CONDITION

As stated earlier, one of the major objectives of the noncompliance and corrective action system is to prevent a repetition of the noncompliance or a similar noncompliance. In order to achieve this objective, the root cause of the noncompliance must be ascertained. Identifying the root cause of a noncompliance means identifying the condition which allowed the noncompliance at hand to occur and which, if corrected, will result in the prevention of future noncompliances identical and similar to the noncompliance at hand. In this regard, a broad perspective must be maintained—broad in both the vertical and the horizontal sense. In the vertical sense, one must understand that the noncompliance at hand exists within a given process and if a repetition of the noncompliance is to be avoided, the condition within that process, which led specifically to the noncompliance, must be identified and corrected. In the horizontal sense it must be recognized that there may

NONCOMPLIANCE AND CORRECTIVE ACTION 129

be similar conditions within other processes which may yield similar noncompliances. These similarities, too, must be identified and corrected. To simply address the singular condition within the process at hand is to take a narrow viewpoint.

The identification of a noncompliance should be viewed as an opportunity to make a safety or technical improvement or a future cost avoidance or both.

The management system procedure should specify the conditions under which the root cause of the problem must be addressed. Specifically, there are four such conditions, as follows:

- When the problem can recur and can adversely impact public and employee safety
- When the problem can recur and any single recurrence will cost considerably more than the cost of identifying and correcting the root cause
- When the problem has been recurring and is expected to continue to recur and the cumulative cost of the future expected recurrences considerably exceeds the cost of identifying and correcting the root cause
- When the problem can recur and its recurrence can cause political or regulatory embarrassment

The procedure should authorize the environmental department to request root cause corrective action under the foregoing conditions, to identify the actionee, to negotiate timely and effective corrective action, to follow-up on the status of the action, and to verify the completion of the action and its effectiveness. The procedure should also address the means by which these things should be documented—using either the environmental open item report or some other form.

The procedure should specify the time limit within which the actionee must submit a response to the environmental department's request for the correction of a problem's root cause. The procedure should allow the initial response to address only the method and schedule for the determination of the root cause, if it is not known. Upon determination of the root cause, a second response should be provided addressing the method and schedule for correcting the root cause. Of course, if the root cause is known at the outset, the initial response should address its correction directly.

The procedure should require the environmental department to issue periodic reports to management similar to the report described in Chapter 11 and depicted in Fig. 8. These reports should provide the status of root cause corrective action with particular emphasis on those cases for which the responses to the root cause request have not been received within the time limit and for which the corrective actions have not been performed in accordance with the commitments.

COMMON MISTAKES IN THE PURSUIT OF ROOT CAUSE CORRECTIVE ACTION

One of the most common mistakes in the pursuit of root cause corrective action is the misapplication of the Pareto principle or, in its more current form, the 80/20 rule. Translated from its original economic origin, the Pareto principle states that a relatively small percentage of causes are at the root of a relatively large percentage of environmental problems, or, stated differently, a relatively large number of environmental problems are attributable to a relatively small number of root causes. The 80/20 rule states that approximately 80 percent of the problems are attributable to approximately twenty percent of the causes.

The Pareto principle and the 80/20 rule should be used in prioritizing the correction of the root causes. In general, given equal safety impacts, the root causes should be addressed in the order of their impact on the company's economics. The Pareto principle does not advocate the disregard of other than the critical few causes. The 80/20 rule does not advocate the disregard of 80 percent of the causes. Often, however, either company resources are inadequate or are misapplied such that root causes other than the critical few are not addressed under the misguided justification of the Pareto principle or the 80/20 rule.

Economics dictate that any action should be taken which provides a yield. If 100 units of cost were to be expended for a return cost avoidance of 200 units (a net avoidance of 100 units) the justification for the expenditure would be clear. However, if 100 units of cost were to be expended for a return avoidance of 110 units (a net return of 10

NONCOMPLIANCE AND CORRECTIVE ACTION

units) the decision is not so clear. In all cases the estimates should be made conservatively. Cost expenditure estimates should be conservative on the high side. Cost avoidance estimates should be conservative on the low side.

In addition, in cases for which public safety is not an issue and economics is, the probability of success also should be taken into consideration. Continuing with the foregoing example, if the expenditure of 100 units has only a 50 percent probability of success in yielding a gross avoidance of 110 units, the real value of the gross avoidance is only 55 units, which compared to the expenditure of 100 units results in a loss of 45 units—certainly not a very good investment. On the other hand, consider a different case in which the expenditure of 100 units can yield an avoidance of 140 units with a 90 percent probability of success, conservatively estimated. In this case the value of the avoidance is 126 units [(0.9)(140)] and, therefore, the expenditure makes economic sense even if the expenditure is for a root cause which is not one of the Pareto critical few or one of the 80/20 twenty percenters.

Another common problem is that there is no procedural definition as to who is authorized to commit an organization to a corrective action and as to the conditions that must prevail as a prerequisite to that commitment. The minimal prerequisites should be:

- Agreement among the affected organizations that the commitment addresses the root cause of the problem
- Agreement that the commitment completion date is both sufficiently timely and achievable
- Understanding and acceptance by both the line management and the assigned individuals of the method by which to achieve the commitment, with individual assignments and dates established for each milepost toward the achievement of the commitment—in other words, understanding and acceptance of the specific plan by which to achieve the commitment. (If the individual contributors who have to get the job done haven't bought into the commitment, how can it be successful?)

Management sometimes imposes commitments or imposes other actions which take away from the resources with which to fulfill earlier commitments and either there is a lack of discipline or a fear of informing management of the consequences of its impositions. Sometimes there needs to be more "give and take" between organizational superiors and subordinates.

Often the system for management follow-up of progress against the mileposts is too gross in that there is no intermediate follow-up between the date on which the commitment is given and the date by which the ultimate action is to be completed. Thus, there is no opportunity to identify the lack of the untimeliness of the action relative to one of the intermediate mileposts along the road to the ultimate action. In such a case, there is no opportunity for remedial action or a mid-course correction, if you will. The ultimate action completion due date may arrive and it may be surprising to find that the ultimate action is incomplete. There should be no surprises.

Sometimes there is no management system procedural definition of the basic elements of information which constitute a commitment and, therefore, commitments are given which are not really commitments in that they are unenforceable. As a minimum, a commitment should comprise statements as to:

- What is to be done
- When it is to be done
- Who is to do it
- What the effectiveness is to be of that which is to be done
- How this effectiveness is to be measured or assessed
- When is this effectiveness to be in evidence such that a state of compliance will be reasonably certain

Sometimes there are no consistent qualitative measures of the frequency with which an organization or an individual is at the root cause of a problem. Similarly, there may be no consistent qualitative measure of the frequency with which an organization or an individual fails to provide a corrective action commitment within the procedurally specified time limit or fails to accomplish the committed work within the commitment due date.

Commitments tracked in one tracking system often are competing with commitments tracked in another tracking system. For example, commitments in response to corporate environmental department audit

NONCOMPLIANCE AND CORRECTIVE ACTION 133

findings being tracked via the audit tracking system may be in competition with commitments to the regulatory agency being tracked via a regulatory commitment tracking system. In such a case, there may be no means by which to prioritize all the commitments relative to one another.

Somewhat as a corollary, managers are frustrated by an inability to have sufficient freedom to choose the things to be corrected. Sometimes environmental department audit finding reports and regulatory inspection reports impose upon the managers a need to address items of marginal or little value simply because the audit and the regulatory tracking systems impose political pressure. Intuitively, the manager may know that his organization's efforts should be directed toward more important things but he has little choice, especially if his resources are unduly constrained or otherwise misapplied as well.

Possibly this next item may leave the reader with the impression of overbearing repetition but, without a doubt, one of the major problems is that corrective actions are not broad enough in scope. Often the same problem must be revisited with new players. Effort is duplicated. To avoid this, in addition to correcting the problem at hand, the process in which the problem existed must often be corrected. Moreover, all other processes in which similar problems exist must often be corrected. Any time there is a problem the basic questions that must be asked are:

- What is to be done to correct the problem at hand? How is it to be done? Who is to do it? When is it to be completed? What are to be the results?
- Until the problem at hand is corrected, is there need for ameliorating or compensating action? If so, what is it to be? Who is to do it? When is it to be done? What is to be its effect? Who is to assure that its effect is the desired effect, recognizing that the ameliorating or compensatory action may temporarily vary from the accustomed procedure?
- What is the root cause of the problem at hand? If the root cause is unknown, what action is necessary to define it? How is the action to be taken? Who is to take it? When will it be completed? If the root cause is known the same questions apply to the actions to eliminate it.

- Does the root cause of the problem at hand exist in other portions of this process or in portions of similar processes? If so, the same what, how, who, and when questions apply.
- Do similar, but not identical, root causes exist in the process at hand and in similar processes? Again, of so, the same what, how, who, and when questions apply.
- Recognizing that the root cause is yet to be corrected, are there any temporary extraordinary actions that should be taken to try to prevent recurrence of the problem until the more technically or economically sound action can be completed, or are there any extraordinary measures that should be applied to assure timely detection of the problem should it recur? If so, the same questions apply.

Until these questions have been asked, answered and acted upon, the corrective action process is incomplete.

Possibly the overriding problem precluding success of the corrective action system could be a lack of a universal sense of a commitment being a promise—a "person's word," so to speak—which may be the case in companies that have not arrived at an overall maturity level aimed at achieving certainty in all matters of compliance. The lack of discipline in one area transgresses to a lack of discipline in another. The failure to take some corrective actions transgresses to the failure to take other corrective actions. Unachieved commitments that are not reflected in an individual's performance ratings lend themselves to still other unachieved commitments. The corrective action discipline must be ingrained in all areas of the company.

13

Contracting for Waste Management Services

INTRODUCTION

Companies which generate hazardous waste have a need to hire contractors to transport, treat, dispose, or store this waste. These contractors will be referred to here generically as "waste management contractors" or simply as "contractors."

The purpose of this chapter is to discuss the conditions under which it makes sense for a hazardous waste generating company to evaluate waste management contractors before awarding a contract, to discuss the objectives and methods of performing the evaluation, and to discuss the considerations in the final contracting phase.

THE NEED TO EVALUATE

A hazardous waste generating company has at least a moral obligation to try to assure that its waste management contractor, in turn, has a program by which to attain compliance with environmental requirements and by which to assure the attainment of compliance. Under certain circumstances the generating company may have a legal requirement as well.

Regardless, the generating company often has a significant economic and public affairs stake in how well the waste management contractor performs his job. Often the waste generating company is much larger than the contractor, and the name of the company is much more well known than the name of the contractor. In the event of a failure on the part of the contractor, the public and the regulatory agencies most likely will seek redress from the generating company as well as from the contractor, regardless of where the fault lies. The news media will focus on the generating company regardless of where the fault lies. The generating company's cost in evaluating potential contractors, in making technically sound contracts, and in auditing to the requirements of those contracts can be much less than the company's cost associated with a single significant incident relative to improper transportation, treatment, disposal, or storage of the hazardous waste.

Therefore, the environmental control management system should require that these contractors be evaluated for environmental factors, that the evaluation precede contract award, and that the evaluation and subsequent contract be such as to provide assurance to the generating company that the contractor, in turn, has reasonable compliance attainment and assurance systems and that these systems are being well implemented. The management system should also assign this responsibility to the environmental department.

PROCUREMENT PACKAGE REQUIREMENTS

The management system should require that the procurement package for any waste management services be processed through the environmental and legal departments.

The environmental department should incorporate requirements into the package which specify:

- The laws governing the waste in question
- Any special inspections and tests of the waste at any stage in its processing, including the inspection and test methods, frequen-

WASTE MANAGEMENT SERVICES

cies and inspection/test personnel qualifications, and including the report formats and frequencies
- The subjects of the contractor's environmental control program procedures which are to be submitted with his bid—subjects such as facility and equipment design review and design change control; equipment maintenance; equipment calibration; routine waste and processing inspections and tests; personnel qualifications, certifications, and training; environmental self-assessment; corrective action; and maintenance of environmental permits and licenses, as a minimum
- The equipment design and performance information which the contractor is to submit with his bid
- The contractor's responsibilities relative to the company's audits and corrective action requests
- The contractor's responsibilities relative to providing facility access to the company's personnel and representatives
- The contractor's responsibilties relative to notifying the company of any noncompliant or otherwise unusual event

The legal department should review these requirements to ensure their legal enforceability. The legal department should also incorporate into the procurement package standard requirements relative to the contractor's insurance and the contractor's indemnification of the company in the event of the contractor's failure to perform in accordance with the requirements of the law or the contract.

PRE-BID EVALUATION

The procedure for contractor evaluations should be structured such as to minimize the environmental department's and company's expenditures while still assuring the performance of technically adequate evaluations. One way of contributing to this objective is for the environmental department to perform pre-bid evaluations—i.e., evaluations of each prospective bidder, using relatively low cost evaluation techniques, as a prerequisite to authorizing that bidder to submit a bid. Through these coarse evaluation techniques obviously unqualified contractors can be removed from the bidders' list, thus precluding the need for full-blown cost and schedule evaluations by other organizations of the company—yielding an overall cost avoidance.

One pre-bid evaluation technique is to request the prospective bidder to submit a written description of his environmental compliance attainment and assurance program. The company's environmental experts may review this program at their home office without incurring the cost of travel to the prospect's facility. The review need not be in depth. Even a cursory review may yield significant findings or sufficient information with which to make a judgment to the effect that a prospect, although marginally adequate in so far as his program is concerned, is significantly below par relative to his competition and, therefore, warrants exclusion from the bidding process.

Another low cost pre-bid evaluation technique is to seek information from other users of the services provided by the prospective contractor, from the regulatory agencies which oversee the contractor's activities and from the community in which the contractor's facilities are located. If these sources do not provide reasons for excluding the contractor from further consideration, they may provide good insight as to weaknesses which should be shorn up as a prerequisite to the contract. Here again, the cost of doing this is relatively low compared to the benefit.

PRE-AWARD EVALUATION

At the conclusion of the pre-bid evaluation process, selected bidders will be invited to bid. The management system procedures should require that each contractor's bid be evaluated for certain nontechnical exclusionary factors before the environmental department and other departments spend resources evaluating the technical portions of these bids. For example, before the environmental department and other departments make any costly assessments of the contractor's technical status, it should be verified that the contractor has an adequate financial status, holds the necessary permits and licenses, and has the required insurance. There is no value contributed in evaluating the technical portions of contractor bids if these contractors are to be excluded from further consideration because of their financial status, the lack of appropriate permits and licenses or the lack of appropriate insurance.

WASTE MANAGEMENT SERVICES

In a similar fashion, the contractor's cost and schedule proposals should be evaluated before the evaluation of their technical proposals. Although a contractor's costs or schedule may not automatically exclude him from further consideration, as in the case of other factors discussed in the preceding paragraph, the cost and schedule evaluation may be used to substantially narrow the field of competitors. Again, it makes little sense for the environmental department and other technical departments to expend resources on contractors whose cost and schedule bids put them out of contention for all practical purposes, recognizing that cost and schedule are given substantial weightings in the overall evaluation process.

Having limited the number of contractors for further considerations, now the environmental department can get to the business of evaluating the technical aspects of the bids. As a minimum, each evaluation should include:

- An assessment of the design capability of the contractor's environmental equipment
- An assessment of the adequacy of his program by which to maintain this equipment
- An assessment of his organization's staffing levels and personnel expertise
- An assessment of his training program
- A detailed assessment of his procedures by which to attain environmental compliance and by which to assure the attainment of compliance
- A detailed assessment of his compliance history

At the point in the process at which a contractor has been selected preliminarily, the environmental department may perform a pre-award survey at the contractor's facility to assess the implementation of the environmental control program.

To help to assure the thoroughness of the evaluation an evaluation checklist should be used, an example of which is provided in Appendix 13.1.

Upon completion of the evaluation and preliminary selection of a contractor and as a prerequisite to the award of the contract, the environmental department should have an objective to negotiate the changes to the contractor's environmental control program which were found to be necessary during the pre-bid and pre-award evaluations. These changes may be nothing more than to provide additional clarity or specificity to existing points in the contractor's environmental control program or they may be to add specific points which are necessary to comply with the law or the procurement requirements. Upon successful negotiations the contract should be accounted for in the environmental department's audit system and the audits should be scheduled.

WASTE MANAGEMENT SERVICES

Appendix 13.1 Hazardous Waste Disposal/Storage/Treatment Facility and Transporter Evaluation Checklist.

I. **Regulatory and Legal Compliance**
   A. *Disposal, Storage, Treatment Facilities*
      1. Has the facility qualified for interim status by:
         a. Being in existence since November 19, 1980?
         b. Submitting a notification form to the EPA?
         c. Submitting a RCRA Part A application to the EPA?
         Provide the facility identification number
      2. Is each type of operation (storage in tank, incineration, etc.) being conducted on site listed in the Part A application?
      3. Has an Act 64 license application been submitted to the MDNR (Michigan only)?
      4. Has the facility been charged by a regulator with a noncompliance of any permit? If "yes," acquire from the facility a separate sheet(s) describing each such noncompliance, its date of occurrence, duration, employee and public health and safety impact, its cause (e.g., inadequate facility/equipment design, inadequate installation/construction, inadequate equipment maintenance procedure, inadequate operating procedure, inadequate administrative procedure, failure to follow procedure, equipment wear-out (specify equipment age), and the status of corrective action acceptance by the regulator.
      5. Has a RCRA Part B application been submitted to EPA? If "yes," answer the following:
         a. Has the permit been issued?
         b. Has the permit been denied or returned? If "yes," acquire from the facility a separate sheet(s) stating the date of denial/return, the reason for the denial/return, the action being taken by the facility and its scheduled completion date.
      6. Has any other environmental permit/license been denied or withdrawn?
      7. Are there any pending or existing regulatory actions against the facility or its parent company which require cleanup of contamination or monitoring of an existing or potential contamination problem? If "yes," acquire from the facility a separate sheet(s) providing the details of each required action.

Appendix 13.1 *(continued)*

   8. Are there any previous, current or pending lawsuits against the facility for environmental damage? If "yes," acquire from the facility a separate sheet(s) providing the details of each.
   9. Have any officers or directors been convicted of criminal violations?
   10. Are the hazardous waste transporters which are used by this facility authorized by EPA to haul hazardous waste?
   If "no," are unauthorized transporters used exclusively to transport small quantity hazardous waste or is characteristic hazardous waste being recycled?
   11. Are the disposal companies used by this facility to handle hazardous waste generated at the facility authorized by EPA to receive and dispose of hazardous waste?

B. *Transporters*
   1. Has the transporter received EPA authorization (receipt of an EPA identification number) to haul hazardous waste?
   If "no," does the transporter exclusively haul one of the following:
   a. Small quantity hazardous waste?
   b. Recyclable small quantity or characteristic hazardous waste?
   c. Nonhazardous liquid industrial waste?
   d. Nonhazardous solid waste?
   e. PCB waste?
   2. Does the transporter have a current MDNR hazardous waste hauler business license (Michigan only)?
   If "no," see item 1, above.
   3. Does each vehicle used to haul regulated hazardous waste have a current MDNR hazardous waste vehicle license (Michigan only)?
   If "no," see item 1, above.
   4. If the transporter hauls only liquids which are small quantity hazardous waste, small quantity or characteristic hazardous waste being sold for recycling, does the transporter have a current Act 136 license (Michigan only)?
   5. Are there any regulatory environmental noncompliance orders issued to transporter? If "yes," acquire the same information as for item I.A.4, above.
   6. Are there any previous, current or pending lawsuits against the

WASTE MANAGEMENT SERVICES

Appendix 13.1 *(continued)*

transporter for hazardous waste contamination or improper disposal? If "yes," acquire from the transporter a separate sheet(s) providing the details of each.
7. Have any officers or directors been convicted of criminal violations?
8. Are disposal, treatment, reclamation, or storage facilities used by the transporter licensed/authorized to receive hazardous waste by Federal and State environmental agencies?

II. **Financial**
   A. *Financial Requirements*
   1. Are the following cost estimates for closure available?
      a. A written estimate of closure costs adjusted annually for inflation?
      b. A written estimate of post closure costs adjusted annually for inflation?
   2. Are the following financial assurances for closure available:
      a. A financial mechanism (e.g., trust fund, demonstration of financial viability) in place to assure the meeting of closure and post closure costs?
      b. A surety bond guaranteeing payment into a closure trust fund?
      c. A surety bond guaranteeing performance of closure?
      d. An irrevocable letter of credit for closure?
   3. Is there a financial test and corporate guarantee for closure?
   4. Does the owner or operator meet the criteria of either "a" or "b," as follows:
      a. Criteria:
         (1) A ratio of total liabilities to net worth less than 2.0
         (2) A ratio of current assets to current liabilities greater than 1.5
         (3) New working capital and tangible net worth each at least six times the sum of the current closure and post closure costs.
         (4) Tangible net worth of at least $10 million.
         (5) Assets in the United States amounting to at least 90% of total assets or at least six times the sum of the current closure and post closure estimates.
      b. Alternative criteria:

Appendix 13.1 *(continued)*

        (1) Current bond rating of AAA, AA, A or BBB issued by Standard and Poor's or Aaa, Aa, A or Baa issued by Moody's.
        (2) Tangible net worth at least six times sum of the current closure and post-closure cost estimates.
        (3) Tangible net worth of at least $10 million.
        (4) Assets in the United States amounting to greater than 90% of total assets or at least six times the sum of closure and post-closure cost estimates.

   B. *Liability Requirements*
     1. Does the owner or operator of the disposal facility demonstrate financial responsibility for sudden accidental occurrences?
     2. Does the owner or operator of the disposal facility demonstrate financial responsibility for nonsudden accidental coverage?
     3. Does the owner or operator of the disposal facility have sudden/accidental occurrence liability insurance in effect which covers at least $1 million per occurrence, and at least $2 million per annum, excluding legal defense costs?
     4. Does the owner or operator of the disposal facility have sudden/accidental occurrence liability insurance which covers as least $3 million per occurrence and at least $6 million per annum, excluding legal defense cost?

III. **General Operations**
   A. *General*
     1. Are there any signs of a recent spill at the facility?
     2. Do the storage areas have secondary containment?
     3. Is there adequate protective clothing and respiratory apparatus?
     4. Are there safety showers and eye washes, and are they properly marked and in the appropriate locations?
     5. Are there warning signs at appropriate places?
     6. Is security adequate?
     7. Is the internal communication system adequate?
     8. Are traffic patterns and routes appropriate?
     9. Is a copy of the emergency plan readily available?
     10. Are there adequate procedures for both personnel and equipment decontamination?
     11. Are there adequate equipment preventive maintenance and calibration procedures?

Appendix 13.1 *(continued)*

    12. Are there adequate training procedures?
    13. Are Material Specification Data Sheets complete, are they readily available, and is there a procedure for their maintenance?
  B. *Hazardous waste storage area*
    1. Is there sufficient space for storage?
    2. Are drums in good condition, with their bungs in place?
    3. Are the barrels placed on pallets?
    4. Is there liquid under the barrels?
    5. Are incompatibles kept separated?
    6. Are appropriate physical barriers used?
    7. Are dikes and berms properly used?
    8. Are liquid storage areas kept clean?
    9. Do containers have stickers stating what they contain and when they were placed in storage?
    10. Are the drums in a proper containment area? Is the base sealed (i.e., concrete with no cracks)? Is the base sloped so that spills will not drain off the base?
    11. Are spills drained into a proper catch basin?
    12. Are tanks covered if kept outside?
    13. Is there a control system for overflow, such as a waste cut-off system or a bypass to another tank?
    14. Is the tank piping or hosing within the containment area?

IV. **Specific Operating Requirements**
  A. *Facility*
    1. General waste analysis:
      a. Is there a detailed chemical and physical analysis of the waste?
      b. Is there a detailed waste analysis plan on file at the facility?
      c. Does the waste analysis plan specify procedures for inspection and analysis of each movement of hazardous waste from off-site?
    2. Do security measures include:
      a. 24-hour surveillance? or
        (1) Artificial or natural barriers around the facility? and
        (2) Controlled access to and egress from the facility?
      b. Danger sign(s) at the facility entrance and the legend in English legible at a distance of 25 feet?
    3. Inspections:
      a. Is the facility inspected for malfunctions, deterioration, oper-

Appendix 13.1 *(continued)*

      ator errors and discharges of hazardous waste that may effect health and safety?
- b. Is there a written inspection schedule at the facility?
- c. If "yes," does the schedule address the inspection of the following items:
  - (1) Monitoring equipment?
  - (2) Safety and emergency equipment?
  - (3) Security devices?
  - (4) Operating and structural equipment (i.e., dikes, pumps, etc.)?
  - (5) Types of problems to be looked at during the inspection (e.g., leaky fitting, defective pump, etc.)?
  - (6) Inspection frequency (based upon the possible deterioration rate of the equipment)?
- d. Are areas subject to spills inspected daily when in use?
- e. Were remedies made for any deterioration or malfunction of equipment or structures resulting from inspection per item c(5), above?
- f. Is an inspection log or summary of inspections for at least 3 years maintained.
- g. Does the inspection log contain the following information:
  - (1) The date and time of the inspection?
  - (2) The name of the inspector?
  - (3) A notation of the observations made?
  - (4) The date and nature of any repairs or remedial actions?

4. Training and records:
   - a. Is there a formalized personnel training program with personnel training records, conducted by a person trained in hazardous waste management procedures?
   - b. Do new employees receive training relative to their needs within six months?
   - c. Do training records indicate that employees' training needs are reviewed annually?
   - d. Do personnel training records include:
     - (1) Job titles?
     - (2) Job descriptions?
     - (3) Descriptions of training received?

WASTE MANAGEMENT SERVICES

Appendix 13.1 *(continued)*

    5. Are the following special requirements for ignitable, reactive, or incompatible wastes addressed:
       a. Special handling?
       b. "No smoking" signs?
       c. Separation and protection from ignition sources?

B. *Preparedness and Prevention*
    1. Is there any evidence of fire, explosion, or release of a hazardous waste or a hazardous waste constituent?
    2. If required, does the facility have the following:
       a. Internal communications or alarm systems?
       b. Telephone or 2-way radios at the scene of operations?
       c. Portable fire extinguishers, fire control equipment, spill control equipment, and decontamination equipment?
       d. An adequate volume of water and/or foam available for fire control? What is the volume?
       e. An adequate check of the water distribution system to assure adequate pressure for fire fighting? Is the check frequency reasonable?
    3. Inspection, testing, and maintenance of emergency equipment:
       a. Have inspection, testing, and maintenance procedures been provided for emergency equipment?
       b. Is the frequency appropriate?
       c. Do the tests cover all of the functional parameters of the equipment?
       d. Is emergency equipment maintained in operable condition?
    4. Has immediate access to internal alarms been provided (if needed)?
    5. Is there adequate aisle space for unobstructed movement?
    6. Have arrangements been made in case of an emergency at the facility with:
       a. Local authorities familiarizing them with the facility operations?
       b. Local hospitals familiarizing them with the types of wastes handled and the illnesses or injuries that could result with improper handling?

C. *Contingency Plan and Emergency Procedures*
    1. Does the contingency plan contain the following information:
       a. The actions facility personnel must take in response to fires,

Appendix 13.1 *(continued)*

   explosions, or any unplanned release of hazardous waste? (If there is a Spill Prevention, Control and Countermeasures (SPCC) Plan, it need only be amended to incorporate hazardous waste management provisions.)
   b. Arrangements committed to by local police departments, fire departments, hospitals, contractors and state and local emergency response teams to coordinate emergency services?
   c. Names, addresses, and phone numbers (office and home) of all persons acting as emergency coordinators?
   d. A list of all emergency equipment at the facility which includes the location and physical description of each item on the list and a brief outline of its capabilities?
   e. An evacuation plan for facility personnel where there is a possibility that evacuation could be necessary? (This plan must describe the signal(s) to be used to begin evacuation, evacuation routes, and alternate evacuation routes?)
2. Are copies of the contingency plan:
   a. Available at the site and local emergency organizations which are committed to provide emergency services?
   b. Available with the Regional Administrator if the facility has submitted a Part B application under RCRA?
3. Emergency Coordinator:
   a. Is the facility emergency coordinator identified?
   b. Is coordinator familiar with all aspects of site operation and emergency procedures?
   c. Does the emergency coordinator have the authority to carry out the contingency plan?
4. Emergency Procedures
   If an emergency situation has occurred in the past:
   a. Were internal facility alarms or communication systems properly activated?
   b. Were appropriate state or local agencies notified on a timely basis?
   c. Were reasonable measures taken to assure that the releases did not spread to other hazardous waste?
   d. Was the character, source, amount, and area extent of the release immediately identified?

Appendix 13.1 *(continued)*

    e. Was treatment, storage, or disposal of the recovered wastes provided immediately after the emergency?

D. *Manifest System, Recordkeeping, and Reporting:*
 1. Use of Manifest System:
    Does the facility have adequate procedures and follow them for processing each manifest, as follows:
    a. Sign and date each manifest certifying that the waste covered by the manifest was received?
    b. Note any significant discrepancies in the manifest?
    c. Immediately give the transporter at least one copy of signed manifest?
    d. Send a copy of the manifest to the generator within 30 days?
    e. Retain a facility copy of each manifest for at least three years from dated delivery?
 2. Manifest Discrepancies:
    a. Are requirements regarding manifest discrepancies met:
       (1) For bulk waste, variations greater than 10% in weight?
       (2) For batch waste any variations in piece count (i.e., one drum in a truckload)?
    b. Are discrepancies resolved within 15 days after receiving waste?
 3. Operating record:
    a. Are written operating records maintained?
    b. Do the operating records contain the following information:
       (1) A description and quantity of each waste received?
       (2) The method (s) and date(s) of each waste's treatment, storage, and disposal?
       (3) The location and quantity of each hazardous waste within the facility?
       (4) A map or diagram of each cell or disposal area showing the location and quantity of each hazardous waste?
       (5) Records and results of all waste analyses, trial tests, monitoring data, and operator inspections?
       (6) Reports detailing all incidents that required implementation of the contingency plan?
       (7) All closure and post closure costs, as applicable?
    c. Are all facility records available for inspection?

Appendix 13.1 *(continued)*

---

    4. Has the facility accepted any hazardous waste from an off-site generator without a manifest or shipping papers?

E. *Monitoring*
   1. Is there a groundwater monitoring system?
      If "no," go to item 2; if "yes," go to item 3.
   2. Is the waiver:
      a. Maintained at the facility?
      b. Certified by a qualified geologist or geotechnical engineer?
      c. Accepted by the regulator?
   3. Does the groundwater monitoring system meet the following requirements:
      a. At least one well installed hydraulically up-gradient from the boundary of the waste management area?
      b. At least three wells installed hydraulically down-gradient at the boundary of the waste management area?
      c. Are the number, locations, and depths of all wells sufficient to yield groundwater samples that are representative of groundwater under the facility?
      d. Are the monitoring wells constructed in accordance with the following:
         (1) Cased (as required)?
         (2) Space between the wells filled with impermeable material?
         (3) Surface/groundwater prevented from entering the well?
   4. Is there a written groundwater sampling and analysis plan that includes procedures and techniques for:
      a. Sample collection?
      b. Sample preservation and shipment?
      c. Analytical procedures?
      d. Chain of custody control?
   5. Is the groundwater sampling and analysis plan followed?
   6. Is the groundwater sampling and analysis plan maintained at the facility?
   7. Is there an *outline* of a comprehensive groundwater quality assessment program that is capable of determining:
      a. Whether hazardous waste or hazardous waste constituents have entered the groundwater?
      b. The rate and extent of migration of hazardous waste or hazardous waste constituents in the groundwater?

Appendix 13.1 *(continued)*

---

    c. The concentration of hazardous waste or hazardous waste constituents in the groundwater?
  8. Has a statistical analysis of the groundwater monitoring data been performed?
  9. Was there a statistically significant pH increase (or decrease) detected in any well?
  10. If "yes," was there a response in accordance with the procedures prescribed in the plan?
  11. Are records maintained and submitted to the regulator?
  12. Is there an annual soil monitoring program?
  13. Is there an ambient air monitoring program?
F. *Closure and Post-Closure*
  1. Closure:
    a. Is there a closure plan for the facility?
    b. Does the plan identify:
      (1) Maximum facility life?
      (2) Maximum hazardous waste inventory?
      (3) Steps to decontaminate equipment?
      (4) Description and schedule of closure activities?
    c. Has the plan been submitted to the regulator?
  2. Post-Closure
    a. Is there a post-closure plan for the facility?
    b. Does this plan contain:
      (1) A description of groundwater monitoring activities and frequencies?
      (2) A description of maintenance activities and frequencies for:
        (a) Integrity of the cap, final cover, or containment structures, where applicable?
        (b) Facility monitoring equipment?
      (3) Name, address, and phone number of the person or office to contact during the post-closure care period?
    c. Has the plan been submitted to the regulator?
G. *Use and Management of Containers*
  1. Are containers in good condition?
  2. Are containers compatible with the waste stored in them?
  3. Do containers meet DOT standards for the type of waste being held?
  4. Are containers stored closed?

Appendix 13.1 *(continued)*

    5. Are containers managed to prevent leaks?
    6. Are containers inspected weekly for leaks and defects?
    7. Are containers stored inside a containment area?
    8. Does the containment area have the following:
        a. An impervious base?
        b. Containment capacity of 110% of the total liquid volume or of the volume of the largest container?
    9. Are ignitable and reactive wastes stored within 50 feet of the facility property line? (Indicate if waste is ignitable or reactive.)
   10. Are all portions of the storage area at least 75 feet from the property line?
   11. Are incompatible wastes stored in separate containers?
   12. Are containers of incompatible waste separated or protected from each other by physical barriers or sufficient distance?
   13. Is the waste handling area constructed so that no waste can escape to groundwater?
   14. Are inventories (receipt and removal) completed twice a month?
   15. Are containers labeled "Hazardous Waste—Federal and State Law Prohibits Improper Disposal"?

H. *Tanks*:
    1. Has the regulator approved the design and construction of tanks?
    2. Are tanks used to store only those wastes which will not cause corrosion, leakage, or premature failure of the tank?
    3. Are tanks equipped with controls (feed cut-off, by-pass systems) to prevent overfilling?
    4. Do uncovered tanks have at least 60 cm (2 feet) of freeboard, or dikes, or other containment structures?
    5. Are waste analyses done before the tanks are used to store a substantially different waste than before?
    6. Are the tanks inspected at least *daily* for discharge control equipment, monitoring equipment (pressure/temperature gauges). and tank level?
    7. Are the tanks inspected at least weekly for leakage and damage to the tanks and the containment system?
    8. Is there a procedure and schedule for completing these inspections?
    9. Are reactive and ignitable wastes in tanks protected from reac-

WASTE MANAGEMENT SERVICES

Appendix 13.1 *(continued)*

        tive or ignition sources or rendered nonreactive or nonignitable prior to being put into the tank?
10. Are incompatible wastes stored in separate tanks?
11. When above ground tanks are located inside a containment structure does the structure have a holding capacity of 150% of the liquids stored inside the largest tank?
12. Do underground tank(s) have a secondary containment and leachate collection and withdrawal system(s)?
13. Is leachate sampling and testing of underground tanks performed at least once a year?
14. Are handling areas constructed to prevent releases of wastes to ground or groundwater?
15. Are tanks located at least 75 feet from the property line?

I. *Surface Impoundments*
1. Does surface impoundment have a liner placed on the bottom and on all earth likely to come in contact with liquids?
2. If "no," has an exemption been granted or an alternate design accepted by the regulator?
3. Are surface impoundments properly designed to prevent overtopping?
4. Do surface impoundments have dikes surrounding the impoundments?
5. Do surface impoundments have at least 60 cm (2 feet) of freeboard?
6. Do earthen dikes have protective covers?
7. Are waste analyses done when the impoundment is used to store a substantially different waste than before?
8. Is the freeboard level inspected at least daily?
9. Are the dikes inspected weekly for evidence of leaks or deterioration?
10. Are reactive and ignitable wastes rendered nonreactive or nonignitable before being stored in a surface impoundment or are they protected from sources of ignition or reaction?
11. Are incompatible wastes stored in different impoundments?
12. Is there a groundwater monitoring program (exception: if the impoundment has a double line and leak detection system)?
13. When leakage has been detected, has the operation of the impoundment been stopped, and has the regulator been notified?

Appendix 13.1 *(continued)*

J. *Waste Piles* (Does not apply if waste will be left on ground at closure)
   1. Are waste piles covered or protected from dispersal by wind?
   2. Has a liner been installed to prevent any migration of wastes out of the pile?
   3. Is each incoming waste or movement analyzed and determined to be compatible before being added to the waste pile?
   4. Are one of the following methods used to assure that no hazardous waste will run off the pile and contaminate the surrounding area:
      a. The pile is on an impermeable base?
      b. There is a run-on control system?
      c. The pile is protected from precipitation or run-on?
   5. Are liquids allowed to be put onto the pile?
   6. Are reactive and ignitable wastes rendered non reactive or nonignitable before being stored in a pile? (Indicate if waste is ignitable or reactive.)
   7. Are incompatible wastes stored in different piles?
   8. Are piles of incompatible waste protected by barriers or distance?
   9. Are waste piles inspected weekly and after storms to detect deterioration, malfunctions, or improper operation run-on, run-off control systems?

K. *Land Treatment*
   1. Are there documented analyses to indicate the biological or chemical degradation capability of the treated waste?
   2. Are run-off and run-on diverted from the facility or collected?
   3. Is wind dispersal of particulate matter controlled?
   4. Is the land treatment area inspected weekly and after storms to detect deterioration, malfunctions, improper operation of the run-on and run-off control systems and improper functioning of wind dispersal control measures.
   5. Are food chain crops prohibited from being grown at the facility?
   6. Is the pH of soil maintained between 6.5 and 9.0?
   7. Is an unsaturated zone monitoring plan available, designed and implemented to detect the vertical migration of hazardous waste and to provide information on the background concentrations of the hazardous waste?
   8. Are records kept regarding application dates and rates, quantities, and locations of all hazardous waste placed in the facility?

WASTE MANAGEMENT SERVICES

Appendix 13.1 *(continued)*

9. Are ignitable and reactive waste not placed in the pile unless measures are taken to assure that the resultant pile is unignitable or nonreactive?
10. Are land treatment zones with incompatible wastes physically separated or managed separately?

L. *Landfills*
1. Does the facility provide the following:
    a. Diversion of run-on away from active portions of the landfill?
    b. Collection of run-off from active portions of the landfill?
    c. Control of wind dispersal of hazardous waste?
2. Does soil underlaying and lateral to active areas have a permeability of 1×10^{-6} cm/sec or less or have a backup timer system?
3. Does the operating record include:
    a. A map showing the exact location and dimensions of each cell?
    b. The contents of each cell and location of each hazardous waste type within each cell?
4. Is landfill inspected at least weekly or after storms to detect: leaks, deterioration in run-off and run-on control systems, proper functioning of the leak detection system, proper functioning of wind dispersal control systems, and the presence of leachate in a proper functioning leachate collection system?
5. Are ignitable or reactive wastes treated so the resulting mixture is no longer ignitable or reactive or such that wastes are disposed of in a way that protects them from any material or conditions which may cause them to ignite?
6. Are incompatible wastes disposed of in separate cells?
7. If bulk or noncontainerized liquids are placed in the landfill:
    a. Does the landfill have a chemically and physically resistant liner system?
    b. Does the landfill have a functional leachate collection system?
    c. Are free liquids stabilized/solidified prior to or immediately after placement in the landfill?
8. Have containers holding free liquids been decanted or have liquids been removed by other methods?

M. *Incinerators*
1. Waste Analysis:
    a. Has an analysis of each waste to be incinerated been performed for its:

Appendix 13.1 *(continued)*

---

      (1) Heating value?
      (2) Halogen content?
      (3) Sulfur content?
      (4) Lead content?
      (5) Mercury content?
  2. Monitoring and inspecting:
    a. Are combustion/emission control instruments monitored at least every 15 minutes?
    b. Is stack plume observed at least hourly for normal color and opacity?
    c. Are the complete unit and associated equipment inspected daily for leaks, spills, and fugitive emissions?
    d. Are emergency shutdown controls and system alarms checked daily for proper operation.

N. *Chemical, Physical, and Biological Treatment*
(Elementary neutralization tanks excepted)
1. Is the material from which the treatment equipment is constructed compatible with the wastes which will be treated so as to not cause equipment leakage, corrosion, or premature failure?
2. Is a continuous feed system equipped with a means of hazardous waste inflow stoppage or control (e.g., cut-off system)?
3. Has waste analysis been performed on each type of waste being treated?
4. Are monitoring equipments, discharge control, and safety equipments inspected at least each operating day?
5. Are the treatment unit and area around the unit inspected for damage or leakage at least weekly?
6. Does the treatment unit have emergency storage capacity to hold all reactants?
7. Are reactive or ignitable waste treated prior to or immediately after placement in the unit?
8. Are incompatible wastes segregated and not placed in the same treatment unit?

14

Maintenance of Environmental Equipment

INTRODUCTION

The purpose of this chapter is to discuss the factors which should be considered in establishing a management system for the maintenance of equipment used to attain and assure compliance with requirements. If this equipment is not well maintained, its service life and reliability may be reduced resulting in the need to take compensatory action by which to attain and assure compliance or, in the absence of compensatory action, the need to derate, curtail, or even stop operations.

It is important to recognize that the success of the total control program is largely dependent, in the final analysis, upon the successful operation of the environmental equipment. In turn, environmental equipment operation is largely dependent upon the adequacy of the maintenance system and its consistent implementation to the environmental equipment. Therefore, it appears appropriate to devote a chapter to the factors to be considered in establishing a maintenance system although the system is applicable to all equipment, not just to environmental equipment.

CHAPTER 14

DEFINITIONS

Equipment is composed of components and parts (items). Each item has design characteristics which may be visual, dimensional, functional, and chemical properties of the item. For example, a visual characteristic of an item may be its paint color. A dimensional characteristic of an item may be its length, width, or thickness between two reference points, or the distance between the center line of a mounting hole and a reference point, or the inside diameter of the hole, or the RMS finish of the inside diameter of the hole. Functional characteristics are such things as tensile strength, yield strength, output voltage, resistance, reliability, and viscosity. Chemical characteristics may be the maximum percentage of carbon in a steel alloy or the maximum allowable pH of a liquid.

Preventive maintenance is applied on a characteristic-by-characteristic basis. Preventive maintenance is applied to the characteristic of the item, not to the item. When preventive maintenance is applied, almost always the characteristic to which it is applied should be in a state of conformance with its requirement, and the purpose of the preventive maintenance for that characteristic should be to optimize the characteristic's condition relative to the requirement. For example, if a surface has become pitted and rusted, albeit within acceptable limits, the pitting and rust should be removed (provided that the minimum wall thickness is maintained); or if the viscosity of the oil has degraded, so that with future use prior to the next maintenance the viscosity will be below its acceptable limit, the oil should be changed; or if the output voltage reading has drifted from the nominal value because of a degraded resistor, the resistor should be replaced and the output voltage reading should be centered on the nominal value.

When preventive maintenance is applied to a characteristic, if the characteristic is not in a state of conformance with its requirement, there may be something wrong with the preventive maintenance itself. As its name implies, the objective of preventive maintenance is to optimize the condition of the characteristic, within reason, so as to try to prevent the nonconformance or failure of the equipment of which the characteristic is a part. If the characteristic's nonconformance or failure occurs before the preventive maintenance action, either the

MAINTENANCE OF ENVIRONMENTAL EQUIPMENT

preventive maintenance may be technically ineffective or its timing may need adjustment. Also, if the nonconforming or failed characteristic is important to the function of the equipment and if the nonconformance or failure is not known until the next preventive maintenance, it could be a serious problem with the design of the equipment in that it does not provide the inherent means by which to detect the nonconformance or failure.

Corrective maintenance, on the other hand, is applied to the characteristic when it is known to be nonconforming or to have failed.

Basically, there are three types of preventive maintenance: scheduled, predictive, and periodic. Scheduled preventive maintenance is based on a knowledge of the characteristic's calendar or operating cycles life. For example, if a specified gasket in a specified application is known to have a life expectancy of at least one year with 95 percent probability, the preventive maintenance for that gasket should be scheduled for one year following its installation and the action should be to replace the gasket. In this case, the gasket's life expectancy is predicated upon calendar time rather than operating time. The schedule for one gasket in one application may be different from the schedule for another, identical gasket in a similar application because the first gasket may have been installed at a different time than the second gasket. For scheduled maintenance there is no need to inspect the condition of the existing, installed gasket. It is to be replaced, even if it looks good.

Predictive preventive maintenance is applied to characteristics for which future noncompliance or failure can be predicted based upon the results of inspections or tests. For example, the amplitude of vibration may be measured, compared to the allowable, and based on that comparison a future schedule for remeasurement may be established. The greater the spread between the actual, measured amplitude and the allowable amplitude, the longer the time before the next measurement need be taken. As the measurements are taken over time, the individual data points will indicate a trend and the slope of that trend will enable the analyst to predict the time at which the actual vibration amplitude will approach the allowable. Obviously, the analyst then will establish a time for a final measurement at which to confirm his prediction and to perform the physical maintenance actions by which to reduce the vibration amplitude to its optimal level—thus restarting the predictive cycle. Predictive preventive maintenance sometimes is also referred to as reliability-centered maintenance.

On the other hand, periodic maintenance requires that a characteristic be inspected or tested at a preestablished periodic frequency and that the characteristic's condition be optimized if the inspection or test results indicate the need to do so. For a critical characteristic, the acceptance criterion should be tighter than given by the design. The reason for this is that periodic maintenance is applied either when there is little knowledge about the life of the characteristic in question or when the characteristic's future state of conformance is not determinable by predictive techniques, as described above. In these cases, for critical characteristics, there is no feasible choice other than to tighten the acceptance criteria for the preventive maintenance inspection or test if the preventive maintenance is to be successful in preventing future nonconformance.

MAINTENANCE SYSTEM CONSIDERATIONS

The management system should require that each item for which preventive maintenance is applicable be identified as such by the design engineering organization at the time that the design is established initially and at any time when the design is modified. As noted earlier, it is insufficient merely to define an item to which preventive maintenance applies. The management system should further require that engineering also define the specific characteristics of each item to which the preventive maintenance applies. For each such characteristic, engineering should specify the type of preventive maintenance to be applied (scheduled, predictive, or periodic). Engineering should specify the initial timing for the preventive maintenance, recognizing that this timing may be adjusted later based on the characteristic's performance results subsequent to the preventive maintenance.

This preventive maintenance information is ancillary design data, which, in accordance with Chapter 5, should be subject to review concurrent with the review of the design itself. The principles of design review discussed in Chapter 5 apply equally here. Obviously,

MAINTENANCE OF ENVIRONMENTAL EQUIPMENT 161

the system should specify the format for this information and the means by which the approved information is to be transmitted for implementation to the maintenance organization.

Each characteristic of an environmental item which is subject to either corrective or preventive maintenance should be reviewed for certain parameters. For example, for each operational output characteristic which can fail or otherwise fault, information must be provided as to how the failure/fault and its correction would be indicated. Common indicators are readouts, recorders, indicating lights, and alarms. This information is a necessary input to the preparation of the operating and maintenance procedures. It is necessary also for repairability. Recognizing that repairability is the probability of detecting, locating, isolating, and correcting a failure/fault and verifying its correction within a specified time limit under specified operating and environmental conditions—how can there be any repairability if the failure/fault is not detectable in the first place? Environmental equipment failures/faults that go undetected can only lead to environmental noncompliance.

For each characteristic subject to either corrective or preventive maintenance, the design should be reviewed to identify any special tools, equipments, procedures, personnel skills, and safety precautions which are needed for the maintenance. Identifying such specialties causes the design engineering organization to reconsider and to justify the design in this regard because these specialties add to the cost of operating and maintaining the equipment. The management system should specify the requirements, responsibilities, and methods by which to provide these specialties when they are justified such that they are available on a timely basis—prior to the time that the initial maintenance may be needed. (Often these same specialties will be needed for the initial fabrication and installation of the equipment as well.)

The system should require that each characteristic subject to maintenance be categorized as to the "organizational level" at which the maintenance is to be performed. For example, assume there to be some sort of an information feedback loop with a sensing device by which to capture a measurement in question, an amplifier by which to amplify the captured measurement signal and a display at which to provide the measurement signal in the control room. Further assume that the amplifier is subject to corrective maintenance because of the potential for failures of the parts within the amplifier. In this example, in the event of an amplifier failure, is the part failure to be located, isolated, corrected, and its correction verified while the amplifier is in its installed position? Or is the amplifier to be removed and replaced, with the failed amplifier being sent to a facility level maintenance shop at which the failure will be located, isolated, corrected, and its correction verified? Or will the failed amplifier be sent to a central corporate-level maintenance shop for its correction? Or will it be sent to the supplier for its correction? Obviously, this information too is necessary for the preparation of the operation and maintenance procedures.

Recognizing the location at which each corrective and preventive maintenance action is to be taken, the system should specify the requirements, responsibilities, and methods by which to provide the appropriate spare or replacement parts at the appropriate maintenance organizational levels. The quantity of spare or replacement parts at each level should be established based on the failure or depletion rate of each part, adjusted for the following factors:

- The criticality of the assembly (e.g., the amplifier) to the attainment or assurance of environmental compliance—the greater the criticality, the greater the number of spare parts needed
- The procurement lead time for the spare or replacement part—the longer the lead time, the greater the number of parts needed
- The commonality of the spare or replacement part—the greater the number of assemblies to which the part applies, the fewer the number of parts needed for each different type of assembly
- The cost of the part—the higher the cost, the fewer the parts desired to be kept in inventory
- The susceptibility of the assembly to design change—the greater the susceptibility, the fewer the parts desired to be kept in inventory for fear of their becoming obsolete
- Whether or not the part has a limited shelf life—the more limited the life, the fewer the parts desired to be kept in inventory for fear of their shelf lives exceeding their limit before they can be used

MAINTENANCE OF ENVIRONMENTAL EQUIPMENT

For low-cost, high-volume parts, provisioning usually is accomplished on a "min-max" basis such that when the minimum quantity of parts in stock is reached, another order is placed to effect an increase up to the maximum quantity. The minimum is calculated to provide adequate assurance that a part always will be available until such time as the reorder is filled. The maximum quantity is calculated to balance the cost of the investment in parts with the cost of repetitive reordering, to arrive at an economic ordering quantity.

The management system should specify the requirements, responsibilities, and methods by which to prepare, review, and approve the operating and maintenance procedures. Among the requirements there should be standards for the format of each type of procedure, for the minimal information to be included in each type, and for the quality of the written and pictorial information.

As part of the review and approval process, the system should require that the operating and maintenance procedural information be verified against its source information, such as the design documents. Also, the system should require that the procedures be subject to a formal change control process such as to assure that any authorized changes (eg., procedural changes that are necessary in view of design modifications) are made on a timely basis and that unauthorized changes are not made. An example of a preventive maintenance procedure for an environmental equipment is provided in Appendix 14.1.

The system should establish the requirements, responsibilities, and methods by which to adjust the spare/replacement parts procurement information based on design modifications. A design modification may introduce new parts or it may make existing parts obsolete, or both.

CALIBRATION

Each environmental equipment which provides a measurement should be subject to calibration by means of a calibration control system. Calibration can be viewed as a form of periodic maintenance. There is no need to go into the details of a calibration control system in that many other books cover the subject adequately.

ENVIRONMENTAL DEPARTMENT RESPONSIBILITIES

Chapter 2 provides a discussion of typical environmental department responsibilities. Among them is the responsibility to assure that adequate procedures exist by which to attain and assure environmental compliance. Operating, maintenance, and calibration procedures are certainly among the types of procedures that are necessary for the attainment and assurance of environmental compliance. Generally, these are areas in which some environmental departments have not played a significant role because environmental equipment is a small portion of all of the equipment which must be operated, maintained, and calibrated, and because the environmental department may not be staffed with environmental equipment operations, maintenance, and calibration expertise. The environmental department should borrow the expertise or hire consulting expertise.

MAINTENANCE OF ENVIRONMENTAL EQUIPMENT

Appendix 14.1 Sample Maintenance Procedure for Environmental Equipment

Environmental Procedure: Precipitator Operation and Preventive Maintenance

1.0 *Purpose*
To provide the responsibilities and requirements for precipitator operation and preventive maintenance.

2.0 *Scope*
This procedure applies to (plant name).

3.0 *Operations Superintendent's Responsibilities and Requirements*
The Operations Superintendent is responsible for the following:
3.1 Once per shift, the Unit Control Operator in the Control Center and the Boiler Operators on the Standard Boilers, shall verify the operating parameters in the Control Center and on the Boiler Panels and shall record the verification of these parameters in the appropriate log book utilizing the rubber stamp format provided per Attachment A. If these parameters are not met, the Shift Operations Supervisor shall be notified and he/she shall take corrective action in accordance with Section 6.0, below.
3.2 Once per day, during the "B" shift, the Precipitator operating parameters and power level conditions shall be checked and recorded on Attachment B, Standard Plant Precipitator Daily Checklist or Attachment C, Reheat Plant Precipitator Daily Checklist. All power level readings shall be made with the field selector switch in the "AUTO" position, if possible. Otherwise, the readings shall be made with the field selection switch in the "MANUAL" position.
Any discrepancies shall be brought to the Shift Operations Supervisor's attention and corrective action shall be taken in accordance with Section 6.0, below.
3.3 A copy of Attachments A, B, and C shall be retained for three years after their issue dates.

4.0 *Technical Superintendent's Responsibilities and Requirements*
The Technical Superintendent is responsible for assuring that the following preventive maintenance activities are performed and that any required corrective actions are taken in accordance with Section 6.0, below.
4.1 Monthly, the Lab shall:

Appendix 14.1 *(continued)*

    4.1.1 Verify that rappers on all units are operating properly and complete the Precipitator/Rapper Report, Attachment D.

    4.1.2 Inspect the ash removal system for proper operation of all valves and controls and for vacuum levels of 11-14 inches of mercury while under normal operation, and prepare a written report of the inspection results.

  4.2 Annually inspect and calibrate each Precipitator Control, and prepare a written report of the inspection/calibration results.

  4.3 Annually inspect T/R Power Control for any unusual problems, check air supply conditions for pneumatic rappers and voltage levels for electromagnetic rappers, and prepare a written report of the results of these inspections.

5.0 *Maintenance Superintendent's Responsibilities and Requirements*

The Maintenance Superintendent is responsible for assuring that the following routine maintenance activities are performed and that any required corrective actions are taken in accordance with Section 6.0, below.

  5.1 During the annual scheduled maintenance outage the following inspections shall be performed and written inspection reports shall be issued.

    5.1.1 Wires and plates for alignment and slack wires;

    5.1.2 Wire support bushings and antisway insulators for general condition, tracking, and cracking;

    5.1.3 Ash hoppers for leakage, ash buildup, and general conditions along the sides, and damage to baffles;

    5.1.4 Insulation on hoppers for damage;

    5.1.5 Plate and wire rappers for proper operation;

    5.1.6 T/R switches for general condition and any possible damage such as oxidation, loose connections, or corrosion;

    5.1.7 T/R oil to assure that it has been tested and that it meets the applicable requirements of the American Society for Testing Materials code for dielectric strength;

    5.1.8 Precipitator expansion joints for leakage and general condition;

    5.1.9 Breakers for general condition and proper operation.

MAINTENANCE OF ENVIRONMENTAL EQUIPMENT

Appendix 14.1 *(continued)*

      5.1.10 Key interlocks for proper lubrication and proper operation.

  5.2 Retain a copy of each inspection report for three years after its issue date.

6.0 *Corrective Action*

Should the above inspections indicate that the precipitator and/or related system are not operating in accordance with established operating parameters, the responsible superintendant shall notify the environmental department representative within 18 hours after the completion of the inspection, and the necessary work shall be performed within a period mutually agreed upon by the Operations Superintendant and the environmental department.

CHAPTER 14

ENVIRONMENTAL PROCEDURE — Attachment A

PRECIPITATOR OPERATION & PREVENTIVE MAINTENANCE

LOG BOOK RUBBER STAMP FORMAT

DATE SHIFT	STANDARD PLANT PRECIPITATOR STATUS
OPERATORS	(ONCE PER SHIFT)
BOILERS FIRED MWLOAD	Y = YES 1 2 3
BFP'S IN SERVICE	N = NO
EQUIPMENT OUT OF SERVICE:	BOILER ON LINE ☐ ☐ ☐
	RAPPER LIGHT ON ☐ ☐ ☐
SOOTBLOWING	RAPPER FLAW ALARM OFF ☐ ☐ ☐
NO. 1 BLR. NO. 2 BLR. NO. 3 BLR.	ALL FIELDS OPERATING IN AUTO POSITION ☐ ☐ ☐
	LIST FIELDS IN MANUAL

```
REHEAT PLANT PRECIPITATOR STATUS
          (ONCE PER SHIFT)
Y=YES  N=NO                         4   5
BOILER ON LINE                     ☐   ☐
MASTER RAPPER CONTROL ON  ☐   ☐
RAPPER LIGHT ON                   ☐   ☐
RAPPER FLAW ALARM OFF        ☐   ☐
ALL FIELDS OPERATING IN AUTO ☐ ☐
LIST FIELDS IN MANUAL
```

MAINTENANCE OF ENVIRONMENTAL EQUIPMENT 169

ENVIRONMENTAL PROCEDURE   Attachment B

PRECIPITATOR OPERATION & PREVENTIVE MAINTENANCE
STANDARD PLANT PRECIPITATOR
DAILY CHECKLIST (B SHIFT)

DATE _____ TIME _____ COMPLETED BY _____

UNIT NO 1

	AUTO OR MAN	PRI AMPS	PRI VOLTS	SEC MA	SPARKS PER MINUTE 0	1-60	OVER 60
A							
B							
C							
D							

STEAM FLOW _____
OPACITY (%) _____
O_2 (%) { E _____ / W _____ }
AIR HTR OUT (°F) _____

UNIT NO 2

	AUTO OR MAN	PRI AMPS	PRI VOLTS	SEC MA	SPARKS PER MINUTE 0	1-60	OVER 60
A							
B							
C							
D							

STEAM FLOW _____
OPACITY (%) _____
O_2 (%) { E _____ / W _____ }
AIR HTR OUT (°F) _____

UNIT NO 3

	AUTO OR MAN	PRI AMPS	PRI VOLTS	SEC MA	SPARKS PER MINUTE 0	1-60	OVER 60
A							
B							
C							
D							

STEAM FLOW _____
OPACITY (%) _____
O_2 (%) { E _____ / W _____ }
AIR HTR OUT (°F) _____

PENTHOUSE
YES = Y   NO = N   NO. 1   NO. 2   NO. 3

AIR LEAKS _ _ _ _ _ _ _ _ _ _ _ _ _ _ _ _ _ _ _ ☐ ☐ ☐
PURGE AIR BLOWER AND HEATER ON _ _ _ _ _ _ ☐ ☐ ☐
RAPPERS WORKING _ _ _ _ _ _ _ _ _ _ _ _ _ _ _ ☐ ☐ ☐
T/R TEMPS. (0-55°) & OIL LEVELS NORMAL _ _ _ _ ☐ ☐ ☐
CONTROL CABINET FANS ON _ _ _ _ _ _ _ _ _ _ ☐ ☐ ☐
COMMENTS _____

HOPPER ROOM
YES = Y   NO = N   NO. 1   NO. 2   NO. 3

AIR LEAKS _ ☐ ☐ ☐
FLYASH REMOVAL VALVES & DIAPHRAGMS OK _ _ _ _ ☐ ☐ ☐
AIR PRESSURE ON FLYASH CONTROL LINES=12 PSI _ ☐ ☐ ☐
FLYASH AIR LINE WATER TRAPS DRAINED _ _ _ _ _ ☐ ☐ ☐
COMMENTS _____

ENVIRONMENTAL PROCEDURE   Attachment C

PRECIPITATOR OPERATION & PREVENTIVE MAINTENANCE
REHEAT PLANT PRECIPITATOR
DAILY CHECKLIST (B SHIFT)

DATE _____ TIME _____ COMPLETED BY _____

STACK OPACITY _____

(SEC MA READINGS FOR A & B FIELDS ON BOTH UNITS TAKEN AT T-R)

UNIT NO 4

GROSS LOAD _____
STEAM FLOW _____
OPACITY (%) _____
O_2 (%) { E _____ / W _____ }
AIR HEATER E _____
OUT TEMP   W _____

	AUTO OR MAN	PRI AMPS	PRI VOLTS	SEC MA	SEC KV	SPARKS PER MINUTE 0	1-60	OVER 60
A				✕				
B				✕				
C								
D								

UNIT NO 5

GROSS LOAD _____
STEAM FLOW _____
OPACITY (%) _____
O_2 (%) { E _____ / W _____ }
AIR HEATER E _____
OUT TEMP   W _____

	AUTO OR MAN	PRI AMPS	PRI VOLTS	SEC MA	SEC KV	SPARKS PER MINUTE 0	1-60	OVER 60
A				✕				
B				✕				
C								
D								

PENTHOUSE
YES = Y NO = N

	NO. 4	NO. 5
AIR LEAKS	☐	☐
PURGE AIR BLOWER & HEATER ON	☐	☐
RAPPERS WORKING	☐	☐
T/R TEMPS. (MAX-55°) & OIL LEVELS NORMAL	☐	☐
CONTROL CABINET FANS ON	☐	☐
AIR REGULATORS ON NO. 5 RAPPERS 30 PSI, WATER TRAPS DRAINED	☐	

COMMENTS _____

HOPPER ROOM
YES = Y NO = N

	NO. 4	NO. 5
AIR LEAKS	☐	☐
FLYASH REMOVAL VALVES & DIAPHRAGMS OK	☐	☐
AIR PRESSURE ON FLYASH CONTROL LINES=12 PSI	☐	☐
FLYASH AIR LINE WATER TRAPS DRAINED	☐	☐
HOPPER HEATERS OVERTEMP LIGHT OFF	☐	

COMMENTS _____

MAINTENANCE OF ENVIRONMENTAL EQUIPMENT 171

ENVIRONMENTAL PROCEDURE   Attachment D

PRECIPITATOR OPERATION & PREVENTIVE MAINTENANCE

PRECIPITATOR/RAPPER REPORT

DATE _____
SR. INITIALS _____
TECH. INITIALS _____

		RAPPER 1-7	RAPPER 8-14	ANNUAL
UNIT #1	"A" FIELD			
	"B" FIELD			
	"C" FIELD			
	"D" FIELD			
UNIT #2	"A" FIELD			
	"B" FIELD			
	"C" FIELD			
	"D" FIELD			
UNIT #3	"A" FIELD			
	"B" FIELD			
	"C" FIELD			
	"D" FIELD			
UNIT #4	"A" FIELD			
	"B" FIELD			
	"C" FIELD			
	"D" FIELD			
UNIT #5	"A" FIELD			
	"B" FIELD			
	"C" FIELD			
	"D" FIELD			

15
Environmental Education and Training

INTRODUCTION

The effectiveness of the total environmental control program is based (1) on the quality of the design of the facility and environmental equipment (e.g., equipment capability, reliability, maintainability, operatibility, etc.), (2) on the quality of the facility and environmental equipment operating systems (i.e., equipment operating policies and procedures), and (3) on the quality of the environmental management systems (i.e., management and administrative policies, requirements, and procedures). The effectiveness of the program is dependent, still further, on the quality of the conformance with these hardware, operating, and management system designs. Finally, conformance with the operating and management system designs is largely a matter of educating and training those who bear the responsibilities for implementing these systems.

Recognizing these things, there is no question that the environmental department should take an active role in assuring the quality of this education and training. The purpose of this chapter, therefore, is to describe the attributes necessary for the quality of the environmental education and training system as it applies to the hardware systems, operating systems, and environmental management systems.

MANAGEMENT FOR ENVIRONMENTAL EDUCATION AND TRAINING

This section addresses the requirements that should be established by means of the management system for environmental education and training (hereafter referred to simply as "training").

The most fundamental requirement of the environmental training system should be that each person who is to perform a procedure that impacts the attainment of an environmental requirement (be it an operating or management systems procedure) should be identified as such, and he should be permitted to perform that procedure only after he has been trained relative to the content of that procedure. For the functions under his control, each supervisor should be responsible for assuring compliance with this constraint.

Either on a centralized basis or on an organization-by-organization basis, the environmental training system should require the supervisor to define the specific operating and management procedures, or portions thereof, for which each person must be trained. For management systems training, the supervisor should recognize that there may be as many as six levels of documents for which a person may require training:

- Environmental policies defining principles of performance which apply universally
- Charters defining organizational responsibilities relative to environmental functions
- Requirements documents defining technical requirements with which the facility must comply.
- Interdepartmental procedures defining requirements, responsibilities, and methods for interfacing among departments in the performance of environmental functions
- Intradepartmental departmental procedures defining requirements, responsibilities, and methods for performing environmental functions within a single organization
- Instructions defining the requirements and methods for performing specialized functions

ENVIRONMENTAL EDUCATION AND TRAINING 175

For many functions, the determination of the specific documents for which a person must be trained should be made on the basis of a task analysis. This analysis is intended to result in an enumeration of each task which must be performed in order to accomplish the function. When the supervisor understands each task, he can better identify any further procedural or skills training which may be required to achieve success in the performance of each task and he can better define the specific content of the training. The environmental training management system should establish the types of functions for which task analysis is required and should establish the method by which to perform the analysis.

The management system next should specify, or require the supervisor to specify in each case, the types of training required. Should it be formal classroom training, on-the-job training, or directed self study? If not preestablished by the system, the supervisor should be responsible for identifying the appropriate contents and sources for the formal classroom training. He should be responsible, also, for establishing the content of the on-the-job and directed self-training.

The system should require the supervisor to schedule the training for each person in his organization. Again, if it has been procedurally preestablished, the system should require the supervisor to make a determination, on a case-by-case basis, as to whether the training is to be delivered only initially or whether it is to be repeated periodically and, if so, its periodicity.

Based on the function's importance to the attainment or assurance of environmental compliance and on the level of environmental safety or financial risk associated with the function, the system should specify the method to be used to assess the individual's capability as a prerequisite to the individual's performing the function. For an important function, the system should require this assessment to be made by formal test and the system should also specify the type of test to be used—an oral test, a written test, a physical capability demonstration test, or some combination of these. If an oral or written test is to be used, the system further should specify the minimum scope of the questions, the method by which the test questions will be secured to prevent falsification of the test results, the minimum passing grade, the remedial action for those who fail, and the number of attempts at passing which may be given to a candidate for the function or position in question. If a demonstration test is to be used, the system further should specify the minimum scope of the demonstrations, the standards with which the demonstrations must comply, the minimum passing grade, the grading method, the remedial action for those who fail the demonstration, and the maximum number of attempts at passing allowed for each candidate.

The system should specify the types of procedural changes for which retraining and retesting is required so as to assure that those who are performing the function which has been changed, fully understand the change and are capable of its implementation.

The system should specify the minimum records to be originated and retained, and their retention periods, by which to demonstrate that the training requirements have been identified, scheduled, and implemented. These records should be on an environmental function-by-function or position-by-position basis. They should identify each person in each function or position and the period during which he performed. The record should show the specific procedural revisions for which each person was trained, the type of training he received, and the date of each type of training. When testing is required, the record should provide the testing data. One should be able to ascertain from the records, on a person-by-person basis, the specific training and tests required, the specific training and test performed and passed, and the training, testing, and assignment dates.

The environmental training management system should require that the environmental department be responsible for reviewing and accepting the training which is established, as a minimum, for the environmentally critical functions and positions—the ones for which tests are required. The environmental department should establish methods by which to assess the effectiveness of the incumbents in these functions and positions. In addition to formal audits performed by the environmental department and self assessments performed by the organization responsible for the function, the environmental department should review environmental noncompliance reports, reports to the environmental regulatory agencies, and operating reports to identify and acquire the correction of any environmental weaknesses in the incumbents or in their training.

ENVIRONMENTAL EDUCATION AND TRAINING 177

Some environmental departments may not be as involved as is necessary in environmental training because the training program for environmental functions should not be much different than the training program for other important functions. Therefore, there may be a tendency for the environmental department to abdicate responsibility for assuring the development and implementation of the training management system which meets environmental needs. Certainly the corporate training organization has its role in that the trainers are expert in the methods of delivering training and in communication skills. This is not enough. The environmental department is expert in the technical area and must exercise its technical expertise as it applies to environmental training.

16

Environmental Awareness and Emergency Response

INTRODUCTION

A major purpose of this chapter is to acquaint the reader with a publication of the Chemical Manufactures Association entitled *Community Awareness and Emergency Response Program Handbook*. This handbood describes the elements of information to be considered for inclusion in the environmental awareness and emergency response plan which should be a part of the company's total environmental control program. Because of the excellence of this handbook, its content is summarized in this chapter with a few recommended additions.

Another purpose of this chapter is to provide a discussion of some factors that should be considered in the mangement system procedure for the development and maintenance of the environmental awareness and emergency response plan. The contents of the plan is one thing; the content of the procedure by which to develop and maintain the plan is something else.

179

PLAN—PLANT EMERGENCY ORGANIZATION

This portion of the plan should designate the individual in charge of each function that must be performed during an emergency situation. These functions may include:

- The determination of the emergency level or status (for example, an emergency alert versus an actual emergency contained within the facility versus an actual emergency that requires notification to community officials outside of the facility but does not require their immediate action versus an actual emergency which requires immediate community official action)
- The making of plant operating decisions at each emergency level
- The responding to and the containment of the emergency itself (for example, in responding to a fire, chemical release, or bomb threat)
- The notifications to company employees, community officials, and officials of neighboring industries which have hazardous material which may be impacted by the emergency
- The communications with the media
- The providing of medical assistance for injured employees and the notifications to their families
- The overall management of the emergency so as to be able to integrate these functions

Of course, for each designee an alternative person should be designated as well. The designees and their alternatives should be specified by name and their work and residence telephone numbers should also be provided.

PLAN—FACILITY RISK EVALUATION

This part of the plan, usually by means of Material Specification Data Sheets, should identify each of the hazardous materials generated or stored at the facility; the quantities of each material, the location of each material, its chemical properties, the symptoms of its adverse effects, and the treatments, including its antidote. The sheets should specify any special handling requirements for each hazardous material.

In addition, this portion of the plan should identify the location of each isolation valve by which to prevent or contain the spread or release of the hazardous material.

PLAN—AREA RISK EVALUATION

This part of the plan should identify other industrial sites within a specified radius of the plant and for each site there should be an enumeration of the types of hazardous material which could be adversely impacted by an emergency at the company's facility for which this plan applies. The names of the persons to be contacted at the neighboring sites, their alternates, and their work and residence telephone numbers should be enumerated. The conditions under which the notifications are to be made should be described.

PLAN—SPECIAL EMERGENCY PROCEDURES

In this part of the plan, the special emergency procedures within the facility should be described. These include the

- Evacuation procedures
- Medical procedures by which to protect against releases or toxic gases
- Special fire-fighting procedures
- Procedures by which to protect against hurricanes, if applicable
- Procedures by which to provide alternate sources of utilities in the event of failures of the primary sources
- Bomb threat procedures

In essence, procedures should be provided to describe the responsibilities and methods for responding to each conceivable type of emergency.

PLAN—EMERGENCY OPERATING PROCEDURES

The procedures by which to operate the facility in an emergency or, as they are called, the emergency operating procedures (EOPs) should be covered in this part of the plan. Each event which conceivably could result in an emergency should be enumerated and analyzed for its credibility. Fault and event trees, as described in Chapter 5, in combination with probabilistic analysis, should be used to determine the event's credibility. For each credible event, given the event scenario and the resultant emergency condition, a detailed, step-by-step procedure should be prepared by which to safely operate the facility or by which to bring the facility's operation to a safe shutdown. Each EOP should consider not only the initiating event and the initial level of the emergency, but also additional credible events and escalating levels of the emergency. The additional events may be secondary failures (i.e., equipment failures that are caused by the initiating failure) or independent failures. The EOP should provide contingency procedures for these additional events. For example, at some point during the emergency, off-site power may be lost due to the growing severity of the emergency or due to the occurrence of a completely independent failure. The EOP should provide the detailed, step-by-step contingency procedure by which to address this potential loss of power. Or, for example, at the start of the emergency, the EOP may provide the operating techniques given the availability of an important component cooling water pump; however, the EOP should also provide the operating techniques should the pump experience a failure. Another way of saying the same thing is that the EOPs should take into consideration both singular credible events and compound credible events which can occur simultaneously or in succession to one another.

PLAN—EMERGENCY COMMUNICATION PROCEDURES

This part of the plan should provide the procedures by which to make each of the following communications or notifications:

- To the employees alerting them to the emergency or to escalating levels of the emergency, directing them to assemble at various locations within the facility or to evacuate any part or all of the facility, and accounting for their presence following an assembly or evacuation
- To the various organizations that are directly involved in fighting the emergency such as fire fighters, rescue teams, and medical personnel both on and off site

ENVIRONMENTAL AWARENESS AND EMERGENCY RESPONSE 183

- To the local officials and response agencies
- To counterparts in neighboring industries which could be affected by the emergency
- To nearby residents
- To corporate management and corporate support organizations
- To regulatory agencies
- To the families of injured employees
- To the news media

PLAN—RETURN TO NORMAL OPERATIONS

For each credible event which could result in an emergency, this part of the plan should specify the minimum conditions which must be satisfied as prerequisites before returning the facility to its normal operations. There should be a direct correlation between the credible events for which there are EOPs and the credible events for which there are prerequisites to returning to normal operations. The plan should specify the responsibilities and methods for meeting these prerequisites, as well as for verifying that they have, in fact, been met.

For each credible event, given that the prerequisites have been verified, there should be a corresponding detailed, step-by-step procedure by which to restart operations.

PLAN—TESTING EMERGENCY EQUIPMENT

In this part of the plan, each type of special emergency equipment should be identified—equipment such as alarms, radios, hotlines, fire fighting apparatus, medical supplies, airborne contamination monitors, fluid leakage monitors, wind and speed indicators, self-contained breathing apparatus, protective clothing, and the emergency control center itself. The required locations for each piece of equipment and their required quantities at each location should be specified. The frequency of the inspection and test of each piece of equipment at each location should be specified, along with the references to the specific inspection and test procedures. The reports to be originated to document the inspection and test results should be specified. Finally, the responsibilities should be specified for each of the following: the performance of the inspections and tests; the assurance that the inspections and tests are performed as scheduled; the preparation of the inspection and test reports; the reviews of the reports to identify and communicate the items which must be repaired or replaced; and, as necessary, the actual repair or replacement and the assurance that it is done correctly.

PLAN—TRAINING

This part of the plan should specify the training requirements, responsibilities, and methods relative to each of the foregoing sections of the plan—i.e., relative to overall emergency organization responsibilities, area and facility risk evaluation, special emergency activities, EOPs, communications, equipment inspection and testing, and drills. The results of which, in large part, indicate the effectiveness of the training efforts. Because Chapter 15 covered the subject of training in detail, nothing more on this subject need be covered at this juncture.

PLAN—EMERGENCY PREPAREDNESS DRILLS

The drills should cover the same areas addressed in the plan and covered by the training.

PREPARATION AND MAINTENANCE OF THE PLAN

The remainder of this chapter addresses the management system considerations for the preparation and maintenance of the plan.

The management system should require the review of design changes or modifications to facilities, equipment or processes which can create, prevent, or ameliorate and emergency and which are needed in response to an emergency.

The management system should also require the review of these modifications for their corresponding impact on the facility and area risk evaluations, special emergency procedures, EOPs, emergency communication procedures, restart procedures, emergency training procedures (including course syllabuses), and emergency drill pro-

ENVIRONMENTAL AWARENESS AND EMERGENCY RESPONSE 185

cedures. The modification of the facility, equipment, or process may impact a simulator or mock-up being used for training. Therefore, the system should also address this possibility. In view of the large number of documents and ancillary equipment which might be impacted by the primary modification, it might be appropriate for the system to provide a checklist to facilitate the completeness of the review process.

The management system should specify the requirements, responsibilities, and methods for performing and documenting these design reviews and the resultant actions and issues, and for assuring that they are brought to a satisfactory resolution. All of the design review management system features described in Chapter 5 apply here equally.

Recognizing that facility, equipment, or process modifications may result in the need for corresponding changes to emergency related documents and ancillary equipment, the management system should specify the requirements, responsibilities, and methods for scheduling and implementing these corresponding changes and for assuring their implementation at the same time at which the basic facility, equipment, or process modification is implemented. It is unacceptable to implement the basic modification and to have the corresponding emergency documentation and ancillary equipment changes lag.

The management system should specify the requirements, responsibilites, and methods for preparing, reviewing, and approving both the area and facility risk evaluations and, in particular, for preparing, reviewing, and approving the Material Specification Data Sheets. The system should require that the sheets be reviewed against facility, equipment, and process modifications which may have introduced new hazardous materials. The system should predesignate the locations to which the sheets should be distributed for easy retrievability, if needed.

The system should specify the requirements, responsibilities, and methods for the preparation, review, and approval of the special emergency procedures and the EOPs. The criteria for their preparation should be specified in detail to provide for consistent procedural format, consistent use of equipment nomenclature, consistent use of action verbs which are specifically defined, and consistent writing clarity and simplicity, as a minimum. The system should specify the requirements, responsibilities, and means by which the procedures will be verified for technical accuracy against their source documents, such as design and technical specification information, and by which the procedures will be validated to assure their practical useability.

The Institute of Nuclear Power Operations in Atlanta has issued EOP preparation guidelines which may be appropriate for application in other industries as well.

Although strict adherence to these procedures is required, the system should specify the level of authority and the conditions which must exist to permit a deviation from a procedure should it be found to be ineffective in an actual emergency situation.

Earlier it was noted that an EOP should be prepared for each "credible" event which could result in an emergency. Certainly the management system should define the criteria by which to make the "credibility" determination.

The system should require that all emergency-related procedures, especially the special procedures and the EOPs in particular, be subject to formal change control by which to assure that unauthorized procedural changes are not made and that the necessary, authorized changes are made. It would be beneficial to be able to directly correlate the procedure revision level to the revision level of the source documents for the procedure. This would help one to determine whether or not revisions to the source documents have been acounted for in revisions to the procedure.

Each procedure subject to validation by means of a drill should be enumerated in the system, in addition to the minimum frequency for conducting each drill. The system should specify the requirements, responsibilities, and method for performing each drill. This includes whether or not the drill will be preannounced, the scope of the drill relative to the scope of the procedure, the means by which performance will be evaluated, and the means by which deficiencies will be identified, reported, and corrected.

Index

Accounting department
  maturity profile and, 9
  tax exemptions and, 84, 85, 86
Air quality, 45
Applications
  defining conditions for, 62–63
  defining source of, 61–62
  examples of, 65–81
    air, 65–67
    general environmental, 80–81
    land, 73–78
    waste disposal, 79–80
    water, 67–72
  origination of, 63–64
Approval applications, 61–81, *see also* Applications
Asbestos, 107

Assessment, *see* Measurement and assessment
Audits, 109–119, *see also* Inspections
  corrective action and, 110, 115, 116–117, 118, 132–133
  education and, 176
  findings of, 110, 115–116, 117
  internal, *see* Self-audits
  maturity profile and, 10
  notification of, 109, 113–115
  planning of, 109, 113–114
  policies on, 22–23, 110, 111, 114, 116
  reporting of, 117–118
  scope of, 109, 110–111, 117
  subject selection in, 109, 112–113

Audits *(continued)*
  team members in, 109, 113, 114
  threats posed by, 115
  waste management services and, 136, 137, 140
Awareness stage, 7, 8–9, 122

Benefit-risk analysis, 24

Cause and effect analysis, 57–58
Certainty stage, 7, 10–11, 12, 122
Certifications, 28
Change control process
  documentation on, 34
  environmental equipment and, 163
Charters, 12, 13, 25
  education in, 174
Chemical Manufacturers Association, 179
Civil penalties, 101
Coal-fixed power plants, 4
Coastal zones, 44
Community Awareness and Emergency Response Program Handbook, 179
Compliance, *see* Noncompliance
Consultations
  in environmental studies, 91, 92
  policies on, 27
Contractors, 63
Cooling water source, 44
Corrective action, 121–134
  audits and, 110, 115, 116–117, 118, 132–133
  documentation on, 33–34

Corrective action *(continued)*
  environmental studies and, 90, 92, 93
  inspections and, 97, 99–102
  maturity profile and, 7, 8, 10
  mistakes in, 130–134
  notification of, 105
  policies on, 27–28, 122, 126
  waste management services and, 137
Corrective maintenance, 126, 159, 161, 162
Cost
  of corrective action, 129, 130–131
  of design engineering, 53, 54
  of environmental studies, 88, 90, 91
  of equipment maintenance, 161, 162, 163
  policies on, 17, 18, 21, 24, 27
  site selection and, 41, 42, 47, 48, 49
  tax exemptions and, 84–85, 86
  of waste management services, 136, 137–138, 139
Credible events, 182, 186
Crosby, Philip, 7

Data collection
  for audits, 109, 114, 115, 117
  for environmental studies, 88, 89, 90
  for site selection, 41, 44–45, 47–48, 50
Dedicated lands, 44, 45
Dedicated waters, 45
Demonstration of compliance, 1–2
Department of Energy, 62

INDEX

Design engineering, 53–60
  equipment maintenance and, 160
  requirements in, 53, 54–56
  review of, *see* Design review
  site selection and, 49
Design review, 56–60
  applied to equipment maintenance, 160
  in emergency response programs, 184, 185
Documentation, 123–126, 128, 129, *see also* Requirements documents
  education in, 174–175, 176
  in emergency response program, 183, 185
  for equipment maintenance, 163
  of inspections, 97, 99, 101
  of noncompliance, *see* under Noncompliance
  nonregulatory, 123
  policies on, 21
  in site selection, 50

80/20 rule, 130, 131
EIS, *see* Environmental Impact Statement
Electric transmissions, 45
Emergency equipment, 182, 183–184, 185
Emergency operating procedures (EOPs), 181–182, 183, 184, 185, 186
Emergency response, 179–186
  area risk evaluation in, 181, 184, 185
  communication in, 182–183, 184

Emergency response *(continued)*
  drills in, 184, 186
  equipment in, 182, 183–184, 185
  hazardous material identification and, 180–181, 185
  individuals in charge of, 180
  operating procedures for, 181–182, 184, 185, 186
  plan preparation and maintenance in, 184–186
  restart procedures and, 183, 184
  special procedures for, 181, 184, 185, 186
  training in, 184
Endangered species, 45
Engineering department
  environmental studies and, 88
  policies for, 24
  tax exemptions and, 84, 86
Enlightenment stage, 7, 9–10, 12, 122
Environmental agencies, *see* Regulatory agencies
Environmental constraint criteria, 45–46, 48
Environmental department
  corrective action and, 132–133
  design engineering and, 60
  documentation and, 32, 33, 34
  educational role of, 173, 176–177
  environmental equipment and, 164
  environmental studies and, 88, 93, 94
  inspections and, 97, 98, 101, 102
  maturity profile and, 9

Environmental department *(continued)*
 policy decisions and, 24, 25–29
 regulatory agency reporting and, 104, 106
 site selection and, 41, 49, 50, 51
 tax exemptions and, 84, 86
 waste management services and, 136
Environmental education, 173–177
Environmental enhancement, 17, 18
Environmental equipment, 157–171, 173
 calibration of, 163
 corrective maintenance for, 126, 159, 161, 162
 design characteristics of, 158–160
 environmental department responsibility for, 164
 failure of, 122–123
 lack of, 104
 maintenance system considerations in, 160–163
 modification of, 23
 preventive maintenance for, 158–160, 161, 162
 procedures for, 163, 165–171
 tax exemptions for, 83–86, 123
 for waste management services, 137, 139
Environmental exclusionary criteria, 44
Environmental Impact Statement (EIS), 2

Environmental laws, 3–4, 123
 application process and, 63
 audits and, 111
 design engineering and, 54
 documentation and, 33
 maturity profile and, 9–10
 policies and, 18, 19, 25
 site selection and, 50, 51
 waste management services and, 136
Environmental performance levels, 31
Environmental Protection Agency, 3, 62
Environmental science, 28–29
Environmental studies, 87–94
 corrective action and, 90, 92, 93
 objective of, 88–89
 plan for, 88
 project manager role in, 88, 92, 93, 94
 report of, 89, 90, 91–92, 93–94
 request for proposal in, 91–92
 schedule of, 88, 90
 scope of, 88, 89
 site selection and, 48
 statement of requirements in, 88, 89–91
EOPs, *see* Emergency operating procedures
Equipment, *see* Emergency equipment; Environmental equipment
Event tree analysis, 58, 182

Failure modes and effects analysis, 58–59

INDEX

Fault tree analysis, 58, 182
Federal Energy Management Administration, 62
Federal Energy Regulatory Commission, 62
Flexibility, 14
Flood plains, 45
Functional role
  audits and, 117
  documentation and, 34
  inspections and, 97, 98, 99, 101
  maturity profile and, 9, 11
  policies for, 27
  reporting and, 104

Geologic anomalies, 44

Hazardous materials, 180–181, 185
Human factors analysis, 60

Initiating event, 58
Insensitivity stage, 7–8, 122
Inspections, 95–102, *see also* Audits
Institute of Nuclear Power Operations, 186
Insurance, 137, 138
Interdepartmental procedures, 12, 13, 14
  education in, 174
Intradepartmental procedures, 12, 14
  education in, 174

Labor availability, 42
Land use, 44
Laws, *see* Environmental laws

Lead time
  in application process, 62, 64
  in equipment maintenance, 162
Leases, 42, 51–52
Legal department, 29
  audits and, 118
  inspections and, 98
  noncompliance notification and, 104
  site selection and, 49
  tax exemptions and, 84, 86
  waste management services and, 136, 137
Licenses, 122
  application for, 61–81, *see also* Applications
  compliance demonstration and, 2
  policies and, 19, 28
  site selection and, 46, 47–50
  waste management services and, 137, 138
Line role
  audits and, 116, 117
  corrective action and, 131
  design review and, 57
  documentation and, 34
  inspections and, 97, 98, 99, 101
  maturity profile and, 9, 10, 11
  policies and, 20, 21, 22–23, 27, 28
  reporting by, 104

Management hierarchy, 12–15
Material Specification Data Sheets, 180, 185
Maturity profile, 7–12, 29, 122
Measurement and assessment
  audits and, 110

Measurement and assessment *(continued)*
  documentation in, 31, 33
  maturity profile and, 7
  policies on, 20, 21, 22, 26–27
Media
  emergency response program and, 180, 183
  environmental studies and, 94
  waste management services and, 136
Michigan Department of Natural Resources, 3
Min-max provisioning, 163

Noncompliance, 1–3
  application process and, 63–64
  audits and, 110, 111, 112, 116
  corrective action for, *see* Corrective action
  design engineering and, 54, 56
  documentation of, 33–34, 123–126, 128, 129
  education and, 175
  environmental equipment and, 157, 161
  factors relating to, 104–106
  inspections and, 98, 99–102
  laws and, 4–6
  management hierarchy and, 14
  maturity profile and, 7–12
  policies to prevent, 19, 20, 21, 22, 23, 24, 26, 27, 28–29
  primary, 125, 126, 128
  reporting of, 103–106, 121–134
    identification and recording in, 123–126
    types of, 122–123
  root causes of, *see* Root causes

Noncompliance *(continued)*
  secondary, 124–125, 128
  site selection and, 42
  waste management services and, 136, 138, 139
Nonregulatory documents, 123

Open item report, 126, 128, 129
Operations department
  documentation and, 34
  environmental studies and, 88
  policies for, 24
  site selection and, 49
  tax exemptions and, 84

Pareto principle, 130, 131
Periodic maintenance, 159, 160
Perk test, 42
Permits
  application for, 61–81, *see also* Applications
  compliance demonstration and, 2
  policies and, 19, 28
  site selection and, 46, 47–50
  waste management services and, 137, 138
Policies, 2–3, 6, 17–29
  on audits, 22–23, 110, 111, 114, 116
  on corrective action, 27–28, 122, 126
  education in, 173, 174
  environmental department role in, 24, 25–29
  management hierarchy and, 12–13
  maturity profile and, 7, 8–9, 10
  planning in, 17, 18

INDEX

Political issues
  corrective action and, 129
  site selection and, 42
Pollutants, 45
Population patterns, 44
Predictive maintenance, 159–160
Preventive maintenance, 158–160, 161, 162, *see also* specific types
Preventive techniques
  audits and, 110
  maturity profile and, 10
Primary event, 58
Primary noncompliance, 125, 126, 128
Probabilistic risk analysis, 58, 182
Procedures, 2–3, 6
  audits and, 110, 111, 114, 116
  corrective action and, 122, 124, 126–128
  education in, 173, 176
  in emergencies, *see* Emergency response
  for equipment maintenance, 163, 165–171
  maturity profile and, 2–3, 6, 7–8
  policies on, 26–27
Project managers, 88, 92, 93, 94
Public affairs department
  audits and, 118
  noncompliance notification and, 104
Public relations
  environmental studies and, 91, 92
  policies on, 17, 18
  site selection and, 42, 50

Public relations *(continued)*
  waste management services and, 136

Quality assurance engineers, 57
Quality controls, 89
Quality Is Free, 7

Redundancy, 58
Regulatory agencies, 3–4
  application process and, 62
  corrective action and, 133
  emergencies and, 183
  environmental studies and, 87–88, 90–91, 92
  inspections by, 95–102
  policies on, 17, 18, 25, 29
  reporting to, 103–107, 125
  site selection and, 48, 49, 50
  tax exemptions and, 85
  waste management services and, 136
Reliability-centered maintenance, *see* Predictive maintenance
Requirements, 2–4, 6
  in design engineering, 53, 54–56
  documentation on, *see* Requirements documents
  education and, 173
  for environmental studies, 88, 89–91
  maturity profile and, 7, 10
  site selection and, 42
Requirements documents, 31–40
  audits and, 110, 111, 114, 116
  examples of, 34–40
  factors to be addressed in, 32–34

Requirements documents *(continued)*
  in noncompliance reporting, 33–34, 122, 124, 125, 126
  reporting and, 104
Restricted use lands, 45
River flow rates, 45
Root causes, 10–11, 121, 128–134
  audits and, 116
  mistakes in correction of, 130–134

Safe-harbor defense, 3
Safety, 127, 129, 131, 175
Samples
  audits and, 114
  inspections and, 97–98, 100
Scenic quality, 45
Scheduled maintenance, 159
Secondary noncompliance, 124–125, 128
Self-audits, 109, 118–119, 176
Site selection, 41–52, 123
  candidate sites in, 44–47
  leasing and, 42, 51–52
  licenses and permits in, 46, 47–50
  potential sites in, 42–44
Specificity
  in application process, 62–63, 64
  in environmental studies, 89, 90
  management hierarchy and, 14–15

Specificity *(continued)*
  in policies, 27
  in site selection, 45
Split sample, 98
Staff role, 9, 10
State designated lands, 45

Task analysis, 175
Tax exemptions, 83–86, 123
Tenets, 1–15
  laws and, 3–4
  management hierarchy and, 12–15
  maturity profile and, 7–12
  total environmental control and, 4–6
Top event, *see* Undesired event
Topography, 44
Total environmental control, 4–6

Undesired event, 57–58

Waste management, 135–156
  evaluation checklist for, 139, 141–156
  pre-award evaluation of, 138–140
  pre-bid evaluation of, 137–138
  procurement package for, 136–137
  site selection and, 44, 45
Water quality, 45
Water sheds, 44
Water withdrawal, 45
Wetlands, 45